NATIONALES
AUTOMUSEUM
THE LOH COLLECTION

FERRARI

ROAD AND RACE MASTERPIECES

JÖRG WALZ

Miles & more

This Mille Miglia marks the final chapter in the annual bitter-sweet madness that has, up until this point in time, enraptured millions of Italians along the roads from Brescia via Rome to Brescia. Ferrari is the brand that is considered the most Italian of all Italian race cars and the marque to beat. On 11 and 12 May 1957, the cars, sporting up to 390 hp, whizzed along the country roads at speeds of up to 300 km/h. Here, photographer Louis Klemantaski, known for his Alfa, Nuvolari and Ferrari pictures, takes a pew in the »hot seat« of Peter Collins in the Ferrari 335 S. Collins/Klemantaski took a commanding lead in the race before retiring approximately 120 kilometres before the finish with a defective differential. What remains are unique testimonies from the Mille Miglia, whose history comes to an end in 1957 with the tragic accident of team-mate Alfonso de Portago

Le Mans sees red

Of the 55 racing cars entering the 24-hour battle in 1960, a dozen hail from Maranello. The factory sent four Testa Rossa cars. Another TR and a phalanx of GTs is entered by privateer customer teams.
In the photo, Wolfgang Graf Berghe von Trips is in the lead ahead of his team-mates Richie Ginther and Ricardo Rodriguez. Von Trips' 250 TR59/60 (9) drops out with an empty fuel tank shortly after this shot. Ginther/Mairesse (10) have to park their Ferrari 250 TRI/60 on lap 204 with gearbox damage. Rodriguez/Pilette finish second in their 250 TR59 entered by NART. Victory goes to their marque colleagues Frère/Gendebien

VOTRE VOITURE
17
10

Marlboro
Bridgestone
AMD
vodafone

The Red Baron

Scuderia Ferrari is the only racing team that has remained loyal to Formula 1 from the very beginning. The Michael Schumacher era stands out in a history rich in glamour. Between 1996 and 2006, »Schumi« stood on the top step of the podium no fewer than 72 times. Between 2000 and 2004, he became world champion five times with the Scuderia.
MSC makes three pit stops in Malaysia in 2004. The seven-time Formula 1 champion also wins this race

CONTENTS

THE FASCINATION WITH FERRARI AT THE NATIONALES AUTOMUSEUM

FOREWORD

For a car museum that wants to show the entire spectrum of automotive history, there is no side-stepping Ferrari. If you are interested in motor racing, there's just no escaping Ferrari's winning legacy. And the moment you mention the word Formula 1, you inevitably end up talking about Ferrari. What's more, the person, Enzo Ferrari, is at least as fascinating as the brand itself, its cars and the people behind the marque. As a schoolboy I was fascinated by the racing driver Wolfgang Graf Berghe von Trips, just as Michael Schumacher has captivated millions and millions of people race weekend after race weekend in the not so distant past.

When composing the exhibition, we focussed on a diverse mix and the fascinating stories behind the individual exhibits. Through these stories the exhibit takes on an extra dimension. The exciting constellation of people and technology makes this dazzling opus of automotive history come alive – and in a unique way.

There are people who move the world. Tactful people who are full of optimism and turn their ideas into reality. This is true in medicine just as in technological development and all other areas. And on a small scale, it also applies to our museum. The Nationales Automuseum offers an unrivalled exhibition; in a very special ambience and with a visitor-oriented presentation format. We want to bring three things together: Automotive history, technology and design. Our aim is to inspire our guests with an enthusiasm for technology.

»The exciting constellation of people and technology makes this dazzling opus of automotive history come alive.«

I am delighted that the Nationales Automuseum has been so well received by its visitors. It fills me with gratitude to see the enthusiasm and beaming faces of the guests as they walk through the exhibition. I know from conversations that our guests leave the exhibition with inspiring impressions.

Our highly motivated team, which brings you closer to the exhibits and the many stories behind them with passion and dedication, certainly contributes to this. I would like to thank the Loh Collection team for this.

The fact that the Nationales Automuseum has been so popular from the outset has also been helped by the broad media coverage, both in this country and in neighbouring countries. Looking at the guest licence plates, Scandinavia and Benelux are just as much a part of it as Spain, France and Italy. The absolute icing on the cake is the award to Nationales Automuseum by the FIA – after just a quarter of a year. But the university co-operations also make it clear to the public that extraordinary things are happening at the museum.

Enzo Ferrari also achieved extraordinary things: he drove his company forward through constant development; he recognised opportunities and knew how to deal with risks. His actions became the DNA of the brand, which has been one of the world's leading luxury brands for decades. Acting in a leading position is not just a matter for an individual or a small team – the entire company has to master it. Enzo Ferrari had this in his blood. It is fascinating to see the speed with which he made fundamental decisions and was always willing to create change and take responsibility for it.

And then, of course, there is the fascination with Michael Schumacher. His personality epitomises passion and ability. He is representative of the »dream team« and its achievements. This era commands my utmost respect; I personally come from a company and have to pay attention every day to what is working and who is not. Much of what Luca Montezemolo, Jean Todt, Ross Brawn, Michael Schumacher and hundreds of other Ferrari 1 team members did can be transferred to our everyday lives...

I am sure that there is something for everyone in our new special exhibition. Everyone will find their favourite story. There are many of them in this book. I am delighted that this exhibition (like our unique Le Mans exhibition) is also accompanied by a very special book. The start of a tradition...
I hope you enjoy reading this book and visiting the Nationales Automuseum – The Loh Collection.

Prof. Dr.-Ing. E.h. Friedhelm Loh

FERRARI AND DESIGN

Ferrari, in essence, is a sentimental marque. Through its passion for motor racing it has been able to create iconic vehicles for the road, whose striking design evokes fascination and enthusiasm throughout the world.

Ferrari's design history is characterised by an evolutionary development of racing and road vehicles that spans more than eight decades.

At the beginning, in the late 1940s and 1950s, form followed function. Ferrari's racing machines were characterised by high-performance elegance and handcrafted details, which were intended to help the cars win. Their mechanics were elegantly disguised, the cars were formally functional and their design focussed on racing performance. The clean lines and characteristic radiator grilles were already highly recognisable Ferrari features that would form the basis for an iconic Ferrari design over the years.

The transition from the racetrack to the road meant that, in addition to a purely functional design language, aesthetic elements also became important in order to make the vehicles attractive to a new clientele, the non-racing drivers. Of course, the racing DNA was to be retained in the vehicles and simple elegance was to characterise the Ferrari design.

As Italy has been a country of beauty and pleasure since ancient times, it is hardly surprising that the first designers also began to design cars at the end of the 19th century. Industrial design already played an important role in the success of products, and so Enzo Ferrari also recognised the importance of vehicle design for his red speedsters. Collaboration with designers such as Michelotti, Vignale and Touring soon followed. However, it was the symbiosis with Pininfarina and his design philosophy of the »couturier of beautiful cars« that inspired Enzo Ferrari, and so Pininfarina became Ferrari's »in-house designer«. In close collaboration with the Ferrari engineers,

»The transition from the racetrack to the road meant that aesthetic elements became important in addition to purely functional ones.«

Wolfgang Henseler is Professor of Digital Media and User Experience at Pforzheim University, Faculty of Design, and Managing Creative Director at SENSORY-MINDS, a design studio for innovative technologies and smart media based in Offenbach am Main. Prof Henseler is responsible for the education and university campus of the »Design« department at the Nationales Automuseum.

the characteristic design features such as the eye-catching radiator grille with the Ferrari emblem, the functionally positioned air intakes and outlets that emphasised aerodynamic efficiency and the powerful lines that gave the vehicles their aesthetic dynamism, began to take shape. These signature design elements found their formal realisation in iconic models such as the Ferrari 250 GT or Ferrari 330 GTO

In the 1960s, sportiness and refined lines increasingly merged with each other, which led to many well-known personalities having their Ferrari designed as a one-off with a special body. The result was modern-day vehicle icons, such as the Ferrari 330 GTC Speciale of Princess Lilian de Réthy from Belgium, which still evoke admiration and fascination for their formal elegance today.

The 1970s at Ferrari were characterised by the design of thoroughbred sports cars, such as the Ferrari 365 GTB/4 Daytona or Ferrari 512 BB. Aerodynamic lines and a striking rear end with round tail lights lent the vehicles their sporty aura and formal fascination.

At the beginning of the 1980s, it was futuristic design elements such as the folding headlights or the striking air intakes on the Ferrari Testarossa that expressed Ferrari's formal design language. At the same time, Ferrari presented the Ferrari 288 GTO, the first super sports car, as a reminiscence of the Ferrari 250 GTO, whose design language was modelled on the Ferrari 512. The Ferrari F40 then showed a formal reorientation. Its more angular design language, the wide rear wing and new materials from motor racing, paired with spartan elegance in the interior and the unmistakable Ferrari colour »Rosso Corsa«, turned the vehicle into a genuine racing car for the road.

In the 1990s, the design language then returned to the lines of the 1970s and the striking wedge-shaped design gave way to softer contours – clearly recognisable in the Ferrari F50.
In 2002, Ferrari introduced a vehicle that was to bear the first name of its founder – the Enzo. The design, which is reminiscent of a fighter jet, places the driver in a road racing car equipped with the latest technology. The upward-swinging doors, the ergonomic cockpit and the steering wheel, which for the first time featured a lot of electronics, paired with significant Ferrari design elements such as the round rear lights that break through the bodywork, made the name »Enzo« a true honour.

From the 2000s onwards, Ferrari pursued a more modern design approach, merging aerodynamic efficiency and sporty styling. The Ferrari La Ferrari and the SF90 Stradale are good examples of vehicles now designed in-house, in which innovative technologies such as LED laser lights give a Ferrari a progressive, sporty appearance.

Conclusio
The design of Ferrari vehicles has always been based on the creative collaboration between engineers and designers. Over the last few decades, it has repeatedly adapted to new circumstances without losing sight of its essential characteristics. From the original racing-influenced design philosophy that form must follow function, to a dynamic elegance with an iconic language of colour and form, Ferrari has timelessly succeeded in combining the myth of fascination with sporting performance paired with intrinsic design. Today, Ferrari continues to facilitate emotions that are the basis of a worldwide fascination with the »prancing horse«.

I wish all readers of this book, and visitors to the Ferrari exhibition at the Nationales Automuseum, a fascinating and insightful experience. The legend that is »Ferrari and design« is, without doubt, an inspirational and unforgettable experience.

Prof. Wolfgang Henseler

»You have to dream a car first.«

QUOTES*

»A Ferrari is a dream – people dream of owning this special car. And for most it will remain a dream, apart from a lucky few.«

»EVERYONE DREAMS OF DRIVING A FERRARI – THAT WAS MY INTENTION FROM THE START.«

»The hippies wanted love and peace. We wanted Ferraris, blonde girls and jackknives.«

~ Alice Cooper

»If you can dream it, you can do it.«

»You have to work continuously, otherwise you have to worry about dying.«

»THE CUSTOMER IS NOT ALWAYS RIGHT.«

»The car was invented to increase people's freedom, not to drive them mad.«

»I don't drive just to get from A to B. I enjoy feeling how the car reacts, becoming part of it.«

* by Enzo Ferrari, unless otherwise indicated

»The most important success is one that must come.«

»I build engines and bolt wheels on them.«

»This is the most beautiful car ever built.«

(On the Jaguar E-Type presented at the 1961 Geneva Motor Show)

»A MAN IS OLD WHEN HE WAKES UP IN THE MORNING WITHOUT A WISH.«

»NO TRIUMPH OR FAME IN THE WORLD IS WORTH AN INCH OF HUMAN SKIN.«

»It takes seven seconds to say that it takes a lot of time to lead Ferrari to the title. But it takes seven years to bring all the elements together.«

~ Jean Todt

»NOBODY REMEMBERS WHO CAME SECOND; AND THAT WILL NEVER BE ME.«

»Racing cars are neither beautiful nor ugly. They become beautiful when they win.«

»The car races of today have hardly anything to do with the sport I have represented for 60 years. It's the fault of the sponsors who have turned the sport into a circus.«

»I don't like being famous – it's like being in prison. And driving for Ferrari would make it even worse.«

~ Valentino Rossi

»WHAT'S BEHIND YOU DOESN'T MATTER.«

»I want to build a car that is faster than all the others. And then I want to die.«

The best driver in the most reliable car with the best tyres — in 2004, the combination of Michael Schumacher, Ferrari F2004 and Bridgestone once again gave the competition a run for their money. Scuderia Ferrari once again dominates Formula 1 at will

FIVE IN ONE GO

Among experts, the F2004 is undoubtedly one of the best F1 cars of all time. And with its impressive track record, #239 is definitely the best of the best. Schumacher won no less than eight GPs with it. A total of nine chassis are created

The F2004 marked the high point of the golden era of the Ferrari »dream team« that Michael Schumacher formed with Ross Brawn and Rory Byrne under the aegis of Jean Todt from 1996 to 2006. According to the official Ferrari version, this Michael Schumacher and Rubens Barrichello company car is the 56th model with which Ferrari has competed in Formula 1. It is the perfect tool for Ferrari's sixth constructors' title and Schumacher's fifth drivers' world championship title under the red banner.

After each of the 18 races, at least one of the two Ferrari drivers finished on the podium. The F2004 has a total of 15 victories to its name. Michael Schumacher achieved 13 of them. A record that will remain valid until 2023, when Red Bull driver Max Verstappen achieves 19 victories in one season (which, however, comprises 23 races). Incidentally, Sebastian Vettel also achieved 13 victories in a total of 19 races in 2013. Over the course of the year, Scuderia Ferrari collected eight double victories and 262 (out of a possible 324) championship points. The duel between »Schumi« and »Rubinho« ended 148:114, with the F2004's only two retirements being accident-related (Schumacher in Monaco and Barrichello in Suzuka).

The incredibly reliable car is so successful that Ferrari also relies on the F2004 for the first two races of the 2005 season; now with the suffix F2004 M (for modificata). A total of five monocoques with chassis numbers 234, 235, 236, 239 and 240 were used on the 18 race weekends. In the following year, Ferrari then used cars 236, 239, 240, 241 and 242. The example in the Nationales Automuseum is the most successful of all F2004s: init Michael Schumacher collects a total of eight victories; seven of them in a row. It bears chassis number 239 and made its debut at the European GP at the Nürburgring.

The brains behind the successes are once again Technical Director Ross Brawn, head Designer and aerodynamics »guru« Rory Byrne as well as chassis chief Aldo Costa and engine man Paolo Martinelli. The F2002 was nicknamed the »red goddess«, whereas the F2004 model was not – perhaps those present at that time had run out of superlatives. The design of the F2004 is based on the previous year's model, the F2003, which in turn goes back to the F2002. The F2004 benefits from new wind tunnel know-how in Maranello. Even engine cooling air and exhaust gases are now directed in such a way that they generate downforce. Compared to its predecessors, the 2004 model can also be tuned more easily to the individual race tracks. Another advantage is the team's extensive testing facilities. With Luca Badoer and Luciano Burti – two F1 professionals under contract as test drivers, and with the in-house circuits of Fiorano and Mugello, Scuderia Ferrari having circuits that can be used all year round, testing is prodigious. The majority of the test kilometres, however, are mostly covered by the work-obsessed Michael Schumacher. The test drives often go on until dusk, so that those present at the track can only see the glowing orange brake disks. The »dream team« leaves nothing to chance...

At that time, Formula 1 was still running on tyres from different manufacturers: Ferrari, Sauber, Minardi and Jordan relied on Bridgestone, while the other teams used Michelin rubber. The contemporary engine regulations at the time required the use of a three-litre, ten-cylinder

In the hands of Michael Schumacher, the F2004 was almost unbeatable at the time. With the car, now in the Loh Collection, Schumacher secured his seventh world championship title

With the F2004 #239, Michael Schumacher collects victories at the Nürburgring, Montréal, Indianapolis, Magny-Cours, Silverstone, Hockenheim, Budapest and Suzuka. A second place behind team-mate Rubens Barrichello comes at Ferrari's home race in Monza

Formula 1 will be using 13-inch rims until the 2021 season. The high tyre sidewalls take on damping tasks. In 2004, both Michelin and Bridgestone are on the grid as tyre suppliers

engine (followed by 2.4-litre V8 engines in 2006). However, while the teams were previously allowed to use an unlimited number of engines, from 2004 only one of the V10 units is permitted per race weekend and vehicle – one of the measures to reduce costs. The teams continue to draw on the full potential when developing the engines: The naturally aspirated engine with a 90° cylinder angle produces around 865 hp (636 kW) at 18,300 rpm in racing mode. It is constantly being further developed and its output is increased to up to 920 horsepower over the course of the season. And such a ten-cylinder engine weighs just 95 kilos.

The minimum weight is also prescribed: 605 kilograms in qualifying, 600 kilograms in all other sessions. The top teams actually undercut this mark; strategically placed

Title-winning workplace of the king — seat-shell and steering wheel are customised for Michael Schumacher

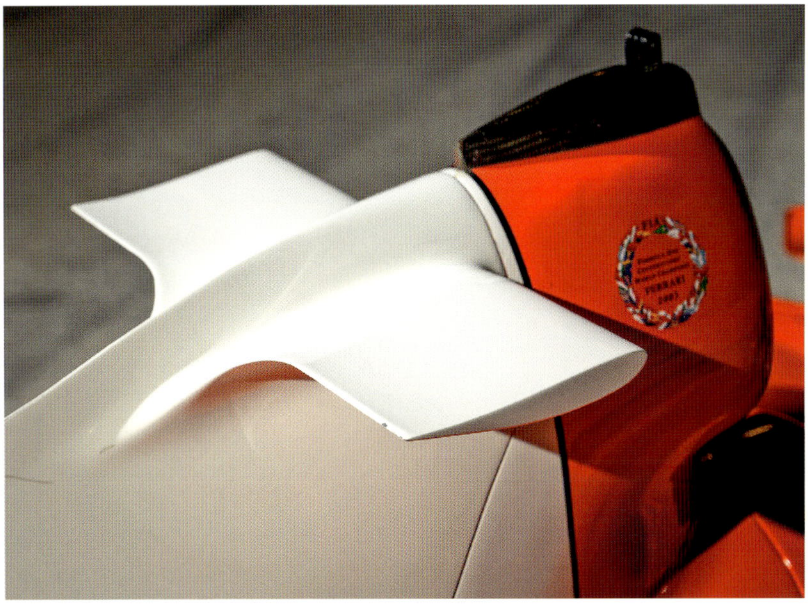

Making good things better — the strengths of the F2004 include not only its reliability, but also the additional aerodynamic fine-tuning, wings, gills and outlets

additional weights ensure compliance with the rules and create an optimum balance of driving behaviour. Pit stops are also obligatory. In 2004, however, not only are new tyres put on the car, but fuel is added to the tank. Accordingly, the designers are able to reduce its volume to a relatively modest 115 litres. During the pit stops, 22 Ferrari mechanics work on the car in a perfectly choreographed manner; up to four times per race and car. The best stops, including refueling, last between 6 and 10 seconds — 12 litres of fuel can be filled per second. Refueling stops have been banned since 2010 for safety reasons.

The cockpit is dominated by a steering wheel that is better described as a command centre: It is peppered with LED indicators, displays, a good dozen buttons, up to ten rotary dials and a stopwatch. There are also four paddles on

»We knew,« says Schumacher, »that we had created one of the most perfect cars of the time. Then we did the first tests and couldn't believe what happened: We beat the Fiorano record by two seconds«

Within a few years, the steering wheels have been transformed into command centres. They are no longer just used to steer and change gear, but also to control countless vehicle functions

In the training workshop, the apprentices at Rittal GmbH & Co. KG built the heaviest F2004 ever from their own scrap, chains and sprockets. During the ten-month construction phase, they repeatedly take measurements at the Loh Collection for the unique two-tonne model

the rear: two for gear selection in the semi-automatic 7-speed gearbox, the other two control the clutch and starting procedure. Drivers use the rotary controls to adjust the stabilisers, differential lock levels, engine mapping and brake balance. There are only two pedals in the footwell above the front axle: for brake and accelerator; with the brakes being operated by the left foot.

Michael Schumacher secures his seventh world championship title ahead of time with second place on his favourite circuit, Spa-Francorchamps. The Scuderia delivers a masterpiece at the French Grand Prix in Magny-Cours, when Schumacher secures victory with a four-stop strategy and puts Renault driver Fernando Alonso in his place. The idea came – during the race – from strategy genius Ross Brawn. Schumacher is made aware of the plan and agrees. The higher number of stops shortens the individual stints. This means that the driver has less fuel in the tank and fresh tyres. But he has to make up for the additional pit stop time. Only drivers who can reliably operate in qualifying mode at any time on command can do this – like Michael Schumacher.

LOTUS 72

The three-litre era also marked the beginning of the triumphant advance of the Ford-Cosworth V8 in 1966, which was used in numerous racing car designs until the mid-1990s. In the second half of the 1960s, aerodynamics also found its way into motorsport: spoilers and wings were added. And in 1968, the rule that racing cars must be painted in the colours of their country of origin was abolished.

The Lotus 72 combines these points in a classic way. The design by Lotus owner, Colin Chapman, was so successful that it was used until 1975. Its first victory dates back to 1970, when Jochen Rindt won in Zandvoort, and the last to 1974, when Ronnie Peterson stood at the top of the podium in Monza. It achieved a total of 20 Grand Prix victories. Jochen Rindt (1970) and Emerson Fittipaldi (1972) become world champions in the Lotus 72 – Rindt posthumously and in the colours of the Gold Leaf cigarette brand, Fittipaldi in the elegant black and gold outfit of John Player Special. Lotus-Ford won the constructors' title a total of three times (1970, 1972, 1973).

The wedge-shaped silhouette of the Lotus 72 established a new design trend. Filigree lightweight construction and detailed technical work remained typical of Chapman's designs. A new feature is the positioning of the radiator to the side of the cockpit. This improves the aerodynamics and shifts the centre of gravity towards the middle of the vehicle. The brakes have been moved inwards at the front and rear in line with the same idea. In addition, this arrangement keeps the braking heat away from the tyres. This allows the Lotus drivers to use softer rubber compounds. Incidentally, treadless slick tyres were used for the first time at the 1971 Spanish GP; another innovation that has become an integral part of racing today.

MCLAREN MP4/5

The 1989 season is characterised by the relentless duel between McLaren team-mates Ayrton Senna and Alain Prost. The previous year, the Brazilian had clinched his first of a total of three world championship titles and was preparing to reach for the crown again. Of the 16 races this season, six will go to Senna and four to the Frenchman, who already has three titles to his name. The decision was made at the finale in Suzuka – in a collision between the two team-mates caused by Prost, in favour of the Frenchman.

One of the strengths of the clearly designed carbon fibre racer is the Honda V10. Its cylinder banks face each other at a 72-degree angle. The naturally aspirated engine revs up to 12,800 rpm and mobilises around 660 hp from a displacement of 3.5 litres. Honda provides eight units of this per GP weekend. In contrast to Ferrari, where a semi-automatic electro-hydraulic gearstick is used for the first time, the six gears of the McLaren gearbox are still shifted manually; corresponding tests with the new gearshift system are not yet promising. On the other hand, the competition has less power; Ferrari is rumoured to have 650 hp and Renault 640 hp.

At the end of the season, McLaren becomes the constructors' world champion. With 141 points, the team from Woking has almost twice as many points on its tally as rival Williams-Renault (77).

The MP4/5, designed by the then McLaren chief designer Neil Oatley, was still in use in 1990, now as the MP4/5B.

Riding the wave of continuing success, the McLaren management around Ron Dennis and Mansour Ojjeh decided to produce a road sports car in the spring of 1989. Under the direction of McLaren mastermind Gordon Murray, the three-seater McLaren F1 is then created.

MERCEDES F1 W07 HYBRID

The W07 is Mercedes' seventh Silver Arrow as a works team. Following joint successes with McLaren, Mercedes opted for its own infrastructure from 2010 onwards. Michael Schumacher even ends his three-year F1 retirement for this. The young Nico Rosberg becomes his team-mate. Another three years later, Schumacher handed over his job to Lewis Hamilton, after three exhausting years of rebuilding. And in 2014 he was finally able to reap the harvest that the team had previously sown. Mercedes also benefits from the introduction of the new 1.6-litre V6 turbo engines with hybrid technology. Mercedes wins 16 of the 19 races in both 2014 and 2015, with Hamilton becoming world champion both times. In the following season, Hamilton and Rosberg fought a fierce duel for the title, with Rosberg also reaching deep into the psychological warfare bag of tricks. It was worth the effort: in the end, Rosberg secured the title.

Incidentally, AMG uses the W07 Hybrid drivetrain as the development basis for the Mercedes-AMG One hyper car, which was delivered from 2023 onwards. In production trim, the two-seater with sophisticated aerodynamics has a system output of 1063 hp (782 kW). Its top speed is electronically limited at 352 km/h.

The F1 team, based in Milton Keynes in the UK, can look back on a chequered history: the Tyrrell years (1968-1998) were followed by the BAR label, the racing team of tobacco giant British American Tobacco (1999-2005). In 2006, the team started as a Honda works team and was transferred into the hands of long-time Ferrari team boss Ross Brawn for the 2009 season. Brawn wins the world championship title with Jenson Button in 2009 and sells the team to Mercedes-Benz. In this way, the Stuttgart-based car manufacturer is able to pick up where it left off in the mid-1950s.

Enzo Ferrari shaped motorsport like no other. With drive and decisiveness, passion and the necessary assertiveness, he shaped the company that has become an integral part of the sport today

CHARACTER AND PASSION

Enzo Ferrari at the wheel of one of his sports cars in the mid-1950s. Lesson learnt – the former racing driver also knows how to drive his vehicles properly

»Racing is an addiction to which you have to sacrifice everything, without exception; without restraint, without hesitation,« says Enzo Ferrari, who achieved influence in the motorsport world like no other in his lifetime. Some people refer to him openly as the »Pope of the North«, or at least as the »King of Maranello«. In 1928, King Vittorio Emanuele III officially awarded him the honorary title of »Commendatore« for his economic achievements. He was also named »Cavaliere« several times (in 1924 and 1925 for his sporting successes, in 1952 for his industrial achievements and again in 1979 for his lifetime achievements as Cavaliere di Gran Croce della Repubblica Italiana). And in April 1960, the University of Bologna made him an honorary engineer (Dr.-Ing. h. c.). »Ingegnere« is Ferrari's favourite of all the titles.

»Ferrari was a typical man of this region,« is how ex-Ferrari sales manager Carlo Tazzioli, who was also born in Modena, tries to describe his former employer. »They say we are people who flare up, but we also cool down quickly. Above all, Ferrari had great charisma. When you were with him, you always had the feeling that you were tiny.«

The people of Emilia-Romagna are considered grumpy but hard-working. The region is prized for the craftsmanship of its inhabitants, especially in the field of metalworking. Enzo's father Alfredo has made a fortune with his own workshop, where up to 30 employees manufacture

Enzo Ferrari at the age of 68 getting into a 330 GT 2+2. The successor model to the 250 GTE was nicknamed »Chinese Eyes« because of its unusual twin headlights

railway parts. Although Enzo had to share a room above the workshop with his brother, who was two years older and also named Alfredo, the craftsman's family did well. As early as 1903, father Ferrari owned a single-cylinder De Dion-Bouton – one of just 27 vehicles in Modena. The family had a chauffeur, and even before the First World War there was a Diatto in the courtyard. His father and brother were car enthusiasts. And when Enzo was ten years old, he was able to experience his first car race. By this time at the latest, the automotive fire was burning in Enzo as well, giving rise to a passion that would ignite countless other automotive passions around the world for decades to come.

Enzo first lost his father in 1916, followed shortly afterwards by his brother. After a short time in the military and in a military hospital, Ferrari looked for a job in the immediate post-war period. He goes to Turin with a letter of recommendation. However, in these very difficult economic times, Fiat had no shortage of applicants. Initially, Ferrari transferred disused Lancia Zeta commercial vehicles, which were fitted with new car bodies, from Turin to Milan for the Giovannoni company. Here he came into contact with the Milanese racing driver circles and became close friends with Ugo Sivocci. He followed him to the car manufacturer CMN (Costruzioni Meccaniche Nazionali S.A.), where new cars were built on the basis of Isotta-Fraschini chassis. In addition to test and transfer runs, Enzo Ferrari also contested his first races with CMN from 1919. In the spring of 1920, together with other companions, Enzo bought an Isotta-Fraschini GP car from 1914 and contested a total of three races in the IR 4500 before joining the Alfa Romeo driver squad in the autumn.

Enzo Ferrari later recalled three career aspirations: opera singer, journalist and racing driver. The first remained out of reach due to his impaired hearing. He pursued the second passion with reports from the motorsport scene, as co-founder of the »Corriere dello Sport« (1924) and through his company's public relations work. The self-made man also fulfilled his desire for a career as a racing driver. But even more than his talent behind the wheel, he was characterised by his gift for making contacts, winning people over, playing them off against each other and motivating them to perform at their best. Enzo Ferrari has a certain feel for favourable opportunities and knows how to exploit them. The young Ferrari went in and out of Alfa in Portello, became general agent for the Emilia region and then, with his Scuderia, became a works racing team. When the Alfa Romeo chapter came to an end in 1939, he used the door that opened at Alberto Ascari's request to launch the Auto Avio-Costruzione project. During the war years, he created jobs away from the front by producing machine tools. And after the end of the war, Ferrari was once again able to gain a foothold in its familiar automotive terrain and in top international sport. Close allies were always to be found at the side of the doer and highly talented networker.

»First of all, developing a new car requires a design. And for a design, you need people who know their way around. Ferrari didn't have a suitable designer, but he knew many people who were prepared to go through fire for him. But what he was looking for was a man of genius, and he found one: Gioacchino Colombo,« recalls Carlo Felice Bianchi Anderloni, head of Carrozzeria Touring. »He was in an unusual position. He was contractually bound to permanent employment at Alfa Romeo, but was not working for the company at the time.« And so Colombo – who had already worked for

Before the Fiorano circuit was built, Ferrari completed its development runs at the Autodromo di Modena. Enzo Ferrari always pays a visit – like here during tests with the 158 before the 1964 season. Incidentally, Enzo won two races here in 1927 and 1928 in an Alfa

Scuderia Ferrari on an Alfa contract before the war – turned his bedroom into a drawing studio. Touring lent him the drawing board on which the design of the first Ferrari was created.

Enzo Ferrari does not accept any conditions. He holds his employees in high esteem, but for him everyone is replaceable. The entrepreneur always had several irons in the fire and alternative options in mind. At the end of the 1940s, Ferrari found himself in the comfortable position of having two highly respected luminaries under contract: Colombo and Lampredi. »He proved to be an experienced exploiter. He picked out engineers and drivers, roped them in, wrung them out and dropped them,« explains chronicler Doug Nye. »But he always had replacements ready beforehand, young lions were always waiting somewhere for their chance should Ferrari's mighty arm strike their superiors. No one was safe.« Throughout his life, Ferrari has known how to stir up discord.

»I have never met anyone as stubborn as myself,« says Enzo Ferrari. »I am driven by a passion that leaves me no time to think or anything else. In fact, I have no interest in a life outside of racing.« As far as his work ethic is concerned, he sets a good example: »He was always in the office for long hours, even at weekends,« his companions recall. »I never really went travelling, never went on holiday. I always spent my best holidays in my workshop – when everyone else was on holiday,« says Enzo Ferrari himself.

The chronicler Pete Lyons describes him as follows: »Enzo Ferrari, a born winner who could raise himself to the status of a titan in a world of giants, knew exactly what he was: a genius. People like that don't care what other people think.« Chris Amon, Ferrari works driver from 1967 to 1969, echoes this sentiment: »He was such an autocrat! You're sitting relaxed in his office – and suddenly Enzo Ferrari jumps up, bursts into wild action, shouts into three telephones at the same time, bangs his fist on the desk like a madman, signal lamps flicker everywhere, alarm bells ring, people rush across the corridors... And suddenly the spook is over again. Enzo leans back in his chair, satisfied and relaxed, grinning a cheeky laugh... He just wanted to show us who's boss here...«

»All the blinds were down in the office. At first I thought it had something to do with the secrecy, but later I realised that Enzo Ferrari liked it dark in the office, a desk lamp was just enough for him,« recalls Gerhard Berger of his first visit to Maranello. »At almost 90 years old, he was by no means the stereotypical dear grandpa, but an ice-cold old man with eyes like a shark (if he ever took off his sunglasses). The first thing he did was pull out a handkerchief and spit into it, then he snorted long and hard. He blew his nose very often and very publicly, it was a part of him.«

Another part of Ferrari's character is his ability to create his own legend. With his proximity to journalists, publications and memoirs, he began to write his own and his company's history early on. Ferrari's press work can at best be compared to that of Porsche, where the importance of PR work was also recognised early on. And Ferrari has a lot to offer and even more to tell. However, a central part of Enzo Ferrari's life is and remains his passion for motor racing: »For me, motor racing is like a woman's room that you dedicate yourself to with passion. But I have never been disappointed by sport, which is more than I can say for people."

HENRY FORD

In a speech by the then US President, to Congress in 2009, Barack Obama unceremoniously attributed the invention of the automobile to the USA. He did not quite hit the bull's eye, but it is thanks to Henry Ford, that the automobile became suitable for mass production. In 1908, Ford transferred the division of labour on the assembly lines of the Detroit slaughterhouses to automobile production. This increased production figures and lowered prices. In this way, he also created a market. Reason enough for many to call him the »Godfather of the automotive industry«.

Henry Ford's vision is a universal low-price vehicle that will put the world on wheels and for which ultimately no factory will be big enough.

Even as a child, the son of Irish immigrants showed considerable skill and began experimenting with combustion engines. Ford worked from 1891 for Thomas Alva Edison, the inventor of the light bulb and electricity. »You've got it figured out,« is Edison's comment when Ford shows him his engine designs. In 1896, Ford presented his Quadricycle. Another three years later, Ford set up his own business. In 1904, he set speed records with his racing car, christened the »Arrow«, which motivated the well-known racing driver Barney Oldfield to set further records with Ford's vehicle all over the country. The car becomes known as the »Oldfield 999«. Ford is also one of the early sponsors of the Indianapolis 500-mile race.

At the centre of his activities, however, was the automobile for everyone. With the Ford Model T – popularly known as the »Tin Lizzie« – he achieved this in an impressive way: Between 1908 and 1927, a total of 16,536,075 units of the Model T were produced, and Ford remains one of the largest and most important automobile companies in the world to this day.

ETTORE BUGATTI

Richard von Frankenberg once called him the »Wizard of Molsheim«. But it was a long road before Bugatti achieved world fame with his car designs. His career began in a bicycle factory. There, at the age of 18, he built his first motorised tricycle. In 1900, Ettore Bugatti built his first car, with which »Bugatti & Gulinelli« won the Gran Premio di Milano in 1901.

With this recommendation, he went to the car manufacturer De Dietrich. In 1904, Bugatti moved on to SACM in Strasbourg and in 1907 to Gasmotorenfabrik Deutz AG in Cologne. At the time, Deutz was regarded as the first port of call for combustion engines, as it was here that Nikolaus Otto, Gottlieb Daimler and Wilhelm Maybach laid the foundations for their triumphant success. By signing up »wunderkind« Bugatti, the Cologne-based company hoped to gain ground and establish itself as an automobile manufacturer. However, Bugattis elaborate design proved to be an economic burden for Deutz. As a result, the two companies parted ways again in 1909. Bugatti moved back to Alsace, to Molsheim, where he set up his own business.

Bugatti had learnt his lesson and now also had an eye on the economic aspects. He started with his team in 1910. His rules were: Don't swim in the current of competition, focus on individuality, but don't follow progress at all costs. The approaches of social reformer John Ruskin were also influential. According to this, an artist should first master his material as a craftsman before designing it. Thus, the aesthetic Bugatti designs are the result of the skills available in Molsheim. After all, the staff of the »Automobiles Ettore Bugatti« were recruited from farm labourers, farmers and winegrowers. But even if there were only a few skilled labourers among them, they made Bugatti the world's top address for automobiles at the time.

FERDINAND PORSCHE

At the age of 13, he began his apprenticeship in his parents' business: a tinsmith's workshop. However, he became interested in electricity. He attended evening classes in electrical engineering. In 1893, he went to Vienna to join the forerunner of the »Unified Electricity Works« and became a guest lecturer at the university.

After working on electric bicycles and carriages, Ferdinand Porsche's early work culminated in an electric car, which he presented for the Lohner brand at the World Exhibition in Paris in April 1900. In the same year, a Lohner-Porsche electric racing car was built, with which Porsche personally competed in the Semmering hill climb. The electric vehicles are followed by hybrids with electric drive and combustion engine at Lohner; so-called »Mixte« designs.

Porsche's career continued via Austro-Daimler to Daimler in Stuttgart. At Austro-Daimler, he developed the »Sascha«. Four of these light vehicles start at the Targa Florio in 1922. Alfred Neubauer – the later Mercedes race director – crossed the finish line in 19th place in one of them after 7 hours and almost 50 minutes. Enzo Ferrari was less than ten minutes faster in his Alfa 20/30 ES, which had a significantly larger engine capacity.

From 1922 onwards, Porsche held the position of chief designer on the board of Daimler. The racing successes of the supercharged models, including victories such as the Targa Florio in 1924, are due to his influence. The S, SS and SSK models bear Porsche's signature.

After a brief period at Steyr, Porsche set up his own design office in 1930. Among other things, the Auto Union Silver Arrow Type C with a V16 mid-engine and the KdF car were developed there. After the war, it would conquer the roads of the world as the Volkswagen.

1863-
1947

1881-
1947

1875-
1951

The history of Ferrari is closely linked to the history of Alfa Romeo. Enzo Ferrari owes much to Alfa Romeo as it has contributed to the success of Enzo's traditional northern Italian brand

QUADRIFOGLIO VERDE AND CAVALLINO RAMPANTE

The Alfa Romeo 16C, a powerful monster with two engines and over 500 hp, is Ferrari's answer to the Silver Arrows from Mercedes-Benz and Auto Union. The »Bimotore« from 1937 is the first vehicle to bear the Cavallino Rampante on the radiator

In the run-up to the 1923 Targa Florio, Ugo Sivocci – Enzo Ferrari's closest friend – adorns his Alfa Romeo RLSS with the Quadrifoglio Verde for the first time. The superstitious Sivocci won and the green four-leaf clover became the mascot for Alfa Romeo's racing cars from then on. According to the legend spread by Ferrari himself, it was the mother of the fighter pilot Francesco Baracca, who died in the First World War, who offered the former mascot to Enzo Ferrari in 1923. This was the rearing horse that would henceforth adorn Ferrari vehicles. Whatever may have motivated the Baracca family to make this gesture five years after the end of the war, today the Cavallino Rampante is the most recognisable trademark in the world. And yet Ferrari was originally unable to have the logo protected: after all, it goes back to the coat of arms of the city of Stuttgart, which the fighter pilot took from one of his 34 downed opponents. It became a symbol for both the Baracca flying squadron and Ferrari – meaning that Porsche and Ferrari ultimately had the same horse in their brand emblems.

Despite the supposed proximity to Porsche, the ties with Alfa Romeo are far closer. Through Ugo Sivocci, Enzo Ferrari entered the Milanese racing circles. The pub, which was frequented by racing drivers and other car people alike, became one of his favourite places. Shortly after Alfa Romeo signed Sivocci as a racing driver in 1920, the

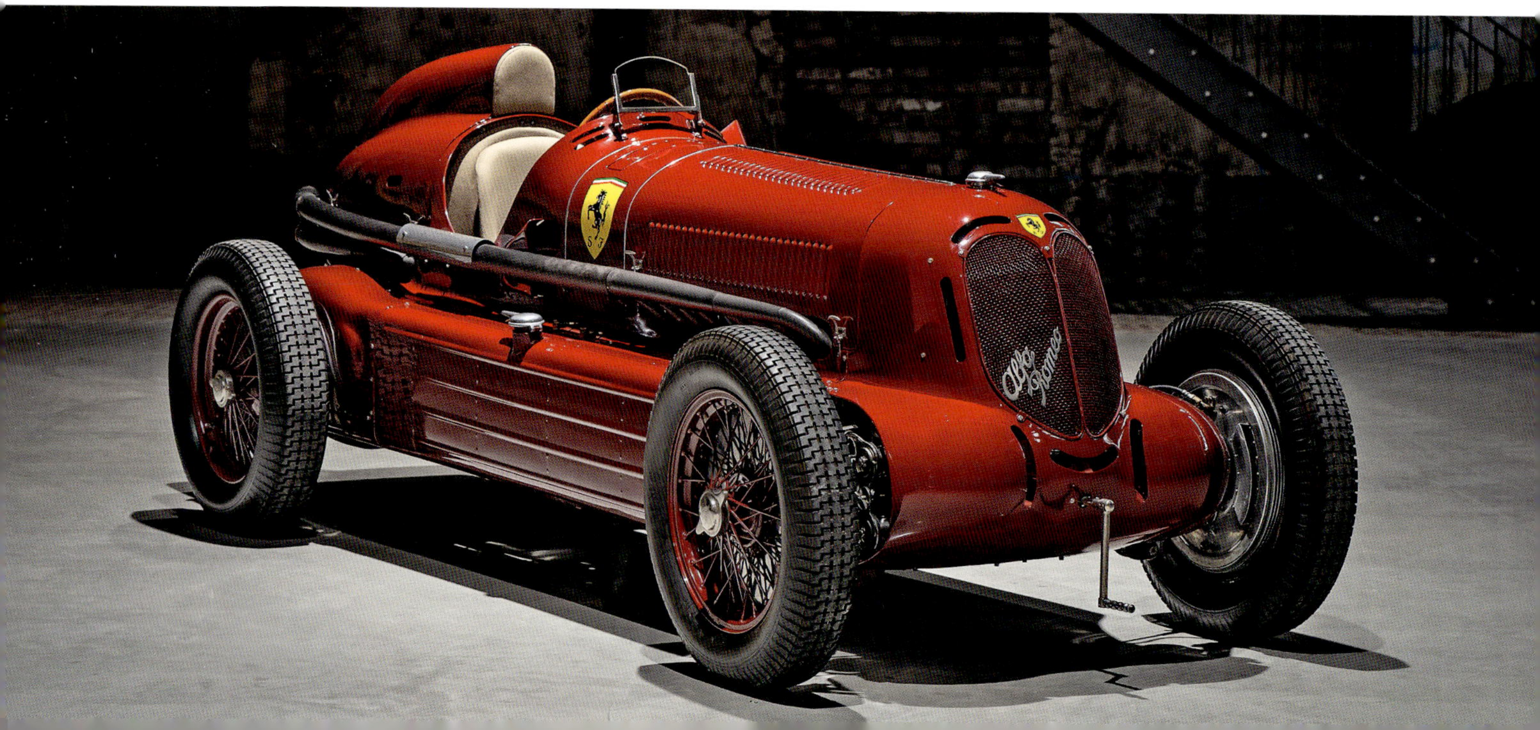

Salerno-born friend, 13 years his senior, brought him onto the team. At the wheel, Enzo Ferrari was considered a good driver, but not a high-flyer. Ferrari contested no fewer than 33 races in the sports and touring cars from Milan up until 1931. These included events such as the Targa Florio, circuit races and hill climbs. Alongside Sivocci, his team-mates were racing personalities of the calibre of Antonio Ascari and Giuseppe Campari. For his second place at the Targa Florio, Ferrari received an incredible 10,000 lire in prize money – a small fortune at the time. Despite several victories, Ferrari was not nominated for Grand Prix races.

And yet he had a major influence on the Milanese brand's GP successes: thanks to Enzo's early talent for bringing the right people together in the right places at the right time, he guided the equally shy and talented constructor Vittorio Jano from Fiat to Alfa Romeo. There, Jano quickly outshone the then chief designer Giuseppe Merosi. With the GP racer P2 and the 6C, designs were created under Jano's direction that established Alfa Romeo's worldwide fame. Enzo Ferrari is in and out of Alfa Romeo. He is a racing driver, headhunter, starts selling Alfa Romeos at his home base in Modena and becomes general representative for the brand in the Emilia, Romagna and Marche regions from 1922. When the Mille Miglia passes through Emilia-Romagna, Ferrari organises the service stations in his region. He always knew how to utilise the potential of his network; for example, the idea of founding a racing team was born at a dinner in 1929. From then on, the Società Anonima Scuderia Ferrari brought together racing professionals, wealthy amateurs and first-class Alfa Romeo equipment.

The money-spinning gentlemen drivers included people such as Carlo Felice Trossi, while the exceptional driving talent Tazio Nuvolari acted as a figurehead. The Alfa works driver and Mille Miglia winner also competed in three races for the young Scuderia Ferrari in 1930. The concept proves successful. Ferrari also quickly finds other solvent sponsors – and thus establishes a liaison with industrial partners that is still fruitful and important in motor racing today. However, these were economically challenging times and the Scuderia also lacked stability. Ferrari's shareholders changed in quick succession.

However, the Modena-based racing team made a name for itself on the international stage and in Italy thanks to its impressive performance and successes. The team became the unofficial works team in the 1932 season, when Alfa Romeo's sales figures – still not a very large company – were weakening. After Enzo Ferrari persuaded Alfa Romeo to release the P3, which had previously been reserved for the factory, the Scuderia was in a better position than ever. However, with financial support from the German Nazi government, the engineers at Auto Union and Mercedes-Benz continued to develop their Silver Arrows. Alfa Romeo sent active support to Modena in order to keep pace. In response to the powerful single-seater from Stuttgart and Zwickau, the 16C, the first vehicle to proudly bear the Cavallino Rampante above the radiator, was created in 1935. The idea for the vehicle came from Enzo Ferrari, the design, under the direction of Vittorio Jano, from Luigi Bazzi, Attilio Marinoni and Arnoldo Roselli. The 16C is a fearsome vehicle: it has two supercharged eight-cylinder in-line engines, hence its second designation: Bimotore (twin engine). One engine is located in front, the other behind the driver. The 3-speed gearbox and differential are interlocked with the front engine. The rear engine can be temporarily decoupled via a clutch. As in the P3, power is transmitted via two V-shaped drive shafts leading to the rear wheels. They rotate under the narrow driver's seat, which is more reminiscent of a camping chair.

Cylinder heads firmly attached to the engine blocks allowed a sensationally high compression ratio of 7:1 for the time (5.75:1 on the normal 8C 2300). At 5400 rpm, the Bimotore delivers 270 hp to each of the two crankshafts. This makes a total of 540 hp. With the 16C weighing just

At the 1951 Swiss Grand Prix, Consalvo Sanesi piloted his Alfetta ahead of Piero Taruffi's Ferrari 375. In the end, Alfa driver Fangio won ahead of Taruffi, followed by three more Alfa 159s. Four races later, Ferrari won a Formula 1 race for the first time, beating Alfa Romeo

At the Nürburgring in 1934: The Mercedes W 25 of Rudolf Caracciola (6) and Luigi Fagioli (9) as well as the Auto Union Type A of the winner-top-be Hans Stuck (1) and August Momberger (2) can be seen in the background. In the foreground are the two Alfa Romeo »Monza« of Attilio Battilana (22) and Ulrich Maag (12) and the Maserati of Goffredo Zehender (20). Scuderia Ferrari will enter the race with three Alfa Romeo Tipo B cars. Scuderia Ferrari driver Tazio Nuvolari will celebrate a sensational victory over the dominant Silver Arrows in one of these cars the following year

1030 kilos, the two in-line eight-cylinder engines have an easy time of it. The independent suspension all round is a novelty. However, the car has a very high centre of gravity and, due to its unbridled power, high tyre wear – at the Tripoli GP, Nuvolari had to pit 14 times to change the tyres. The two engines are also thirsty. As a result, the 240 litres of fuel they carry weigh heavily. Scuderia Ferrari produces two vehicles for 1935. Nuvolari came second on the Berlin AVUS high-speed circuit. However, the »flying Mantuan« insisted Enzo Ferrari recommend that the Bimotore be retired in favour of the P3. He was proved right, as demonstrated by his victory at the German GP that same year.

To prevent the powerhouse from being quietly and secretly sidelined, Scuderia Ferrari undertook record runs on the motorway from Florence to Livorno. The sports stewards recorded a top speed of 323 km/h. One of the two Bimotors survives. Today it is part of the Loh Collection – the factory's own Museo storico Alfa Romeo has a recreation built in the 1970s based on the design plans. Despite the further development of the Tipo B (P3) and other models, such as the Tipo C (8C, 1935, and 12C, 1936), the oxblood-red painted Alfa works cars of the Scuderia Ferrari lost out to the silver arrows.

In 1938, at the behest of Italian fascist leader Benito Mussolini and under the direction of Alfa boss Ugo Gobbato, Alfa Romeo decided to enter the race again with a factory line-up orchestrated in Milan. Enzo Ferrari becomes racing director of the Alfa Corse works racing team. And wealthy: the Milan-based company buys 80 per cent of his Scuderia. At that time, the SF engineers were well advanced with their project for a racing car with a 1.5-litre eight-cylinder supercharged engine. Enzo Ferrari takes most of his employees, the machinery, the old cars, all spare parts and four brand new »Alfetta« 158s to Portello. What remains on Viale Trento e Trieste in Modena is the Alfa branch of Enzo Ferrari. In view of the success of the Auto-Union mid-engined racers, Ferrari also considered realising the Vetturetta 158 in a mid-engined design. Enzo Ferrari succinctly rejects this idea: »It has always been the case that the ox pulled the cart.«

In Portello, the successful and enterprising entrepreneur Enzo Ferrari soon realised that he can no longer be integrated into corporate structures. He constantly clashed with Wifredo Ricart, the head of the development department, who had been brought to Alfa from Spain. And so, in 1939, Alfa Romeo and Enzo Ferrari parted ways, with the latter using the severance payment he received as an investment for future projects. This was followed – despite the non-competition clause – by the brief chapter of Auto Avio Costruzione, machine tool production during the war years, and a new start with the production of racing and sports cars, which henceforth bore the Ferrari name. After the war, Alfa Romeo entered the race with the Alfetta racing cars, which were created at Ferrari in 1937. Their design proved to be so superior that the Milan-based company won the Formula 1 titles in 1950 and 1951. However, after an initial learning curve, Ferrari also played its part in the premier class: Scuderia Ferrari achieved its first F1 victory at Silverstone in 1951. Not without the typical theatrics of Enzo Ferrari, the »Commendatore« remarks: »I cried with happiness. But my tears of joy were also mixed with pain, because I remembered that I had killed my mother that day.« The Ferrari victory ended the dominance of the previously unbeatable Alfa Romeo. And it marked the beginning of a success story in which Ferrari would collect 243 F1 victories, 16 constructors' titles and 15 drivers' titles (as of 2023). This success was also due to Enzo Ferrari's uncompromising approach to personnel issues: while the first Ferrari racing cars were created under the direction of Gioacchino Colombo (who also penned the Alfetta), the successful 1951 F1 engine bore the signature of Aurelio Lampredi, who was spontaneously promoted to Ferrari chief designer. Colombo was quickly demoted and allowed to focus on sports car development; he moved back to Milan in 1951.

ALFA ROMEO 8C 2300 »MONZA«

After Enzo Ferrari arranged the transfer of Vittorio Jano from Turin to Portello, Alfa Romeo produces two extremely successful vehicles: the Grand Prix racer P2 and the 6C 1500/6C 1750. Jano's next coup followed in 1931 with the 8C 2300. Versions with a long and short wheelbase were produced for this sporty car. Different bodies from various coachbuilders were built on it. There were coupés, saloons, spiders and cabriolets. They are suitable for a trip to the theatre, but also for participation in touring car, sports car or even Grand Prix races.

In the 1930s, Carrozzeria Touring enjoyed the status of house and court supplier to Alfa Romeo. However, the Carrozzeria Zagato and Zagato-Interims successor Brianza were also consulted for the lightweight open-top competition cars.

The sports cars from the pen of Vittorio Jano were the favoured sports cars of Scuderia Ferrari, which began racing in 1929 with a series of 6Cs. These were soon replaced by the 8C models. Its supercharged in-line eight-cylinder engine produced 142 hp in standard trim. The competition cars entered by Ferrari achieved up to 165 hp. A lightweight two-seater with revised suspension and just under 180 hp was created for the Italian GP.

Six of these cars were built in 1931 and a further four in 1932. In Monza, they entered the race without mudguards and lights. However, these components could easily be refitted for other events. As a reminiscence of its first (and successful) use, the most powerful 8C variant was given the name »Monza«. For the 1933 season, Scuderia Ferrari redesigned six of these vehicles, which could reach speeds of up to 225 km/h, and increased the engine capacity to 2.6 litres.

ALFA ROMEO P3 (TIPO B)

The Tipo B is the first monoposto to make its debut in GP racing. This marked the beginning of a tradition of single-seater formula racing cars that continues to this day. Until the 750-kilogram formula came into force in 1934, the P3 (as it was also known) dominated the races in which it competed.

It was powered by a further development of the 2.3-litre eight-cylinder engine, which initially had a displacement of 2.6 litres in the P3, later increasing to 2.9 litres. The output of the engine, which was supercharged with two Roots compressors, increased by almost 20 percent from 215 to 255 hp.

When Alfa Romeo completely outsourced its racing activities to Scuderia Ferrari in 1933, the foundations were also transferred to Modena shortly afterwards. The reason for Alfa's withdrawal was economic problems in Milan. These ultimately lead to Alfa Romeo being incorporated into a state holding company from 1933. Meanwhile, Ferrari allowed the masterpieces from Milan to continue to shine on the sporting stage. An aerodynamically panelled version was also created for the AVUS race in 1934. Guy Moll wins in Berlin with the P3 »Aerodinamica«, which is formed according to the plans of aeronautical engineer Cesare Pallavicino and whose top speed is 20 km/h higher than that of the »naked« P3. Another triumph on German soil becomes legendary: At the Nürburgring, Tazio Nuvolari outdid the assembled Silver Arrow competition in 1935. It is probably the most significant victory in the impressive track record of this Jano design.

The open design of the P3 allows interesting insights: under the Spartan driver's seat, for example, the centrally positioned differential can be seen, from which the two drive shafts branch off in a V-shape towards the rear axle.

ALFA ROMEO 158 (ALFETTA)

While the Silver Arrows from Mercedes-Benz and Auto Union dominated Grand Prix racing in the pre-war years, manufacturers such as Maserati, Bugatti and Alfa Romeo also focussed on smaller vehicles based on the Voiturette regulations. This included racing cars with a maximum engine capacity of 1.5 litres.

While Scuderia Ferrari tried its hand more or less unsuccessfully at GP racing cars with 8, 10, 12 and 16 cylinders, Enzo Ferrari fired the starting signal for the small Alfa Romeo, the »Alfetta«, in Modena, in 1937. The young engineer Gioacchino Colombo was responsible for the design. He was head of the special designs department in Portello and was sent to Scuderia Ferrari. There, under his direction, a vehicle was created that was subscribed to victories right into the Formula 1 era.

Colombo learnt his trade under Vittorio Jano, who moved to Lancia in 1938. Alberto Massimino (who designed the gearbox and rear axle), Angelo Nasi and Federico Giberti (front axle and steering) also worked on the small Alfa. The later Alfa chronicler Luigi Fusi also contributed to the Tipo 158: he transported the drawings and calculations from Portello to Modena. The type designation 158 reveals the displacement and number of cylinders of the supercharged single-seater, which sensationally debuted in August 1938 at the Coppa Ciano Junior in Livorno with a commanding victory.

From then on, however, the Tipo 158s entered in the race sporting the colours of Alfa Corse – Alfa Romeo dissolved Scuderia Ferrari at the beginning of 1938 and appointed Enzo Ferrari as racing manager of the works team.

They were used again from 1946 and were further developed into the Tipo 159 in 1951. Farina and Fangio became the first Formula 1 champions with the Alfetta.

1931

1932

1938

HOW TIME FLIES

In the 1950s, the Indy 500 is part of the F1 World Championship. But the US teams kept to themselves. Ferrari's start in 1952 is an exception. Alberto Ascari's race ends – after leading for miles – with wheel bearing damage

(1947 – 1956)

The first decade

Team boss and entrepreneur Enzo Ferrari had long since set the course. The site and property on Via Abetone in Maranello were already occupied during the war, and the charismatic leader already knew that his companions were at work. The first test drive of the premiere model took place in March 1947. It was given the type designation 125, and with the first victory of the car, which now bore the name of the company founder, the heart of the Ferrari brand really began to beat.

It all began with success at the Rome Grand Prix. With his victory, Franco Cortese occupied the position in which Ferrari drivers would henceforth be encountered in motorsport on all continents; at the front is where Ferrari is to be found. With its first victory in the Mille Miglia, Ferrari leaves a red mark in the annals of the »Mille«, winning the road racing classic eight times until its end in 1957. Nine victories at Le Mans between 1949 and 1965. Triumph number ten in 2023. Ferrari is also successful at the Carrera Panamericana in Mexico.

The Le Mans victory in 1949 marked the start of internationally recognised successes for Ferrari

With the Barchetta 166 MM, Ferrari collected countless class victories and almost five dozen overall victories in prominent events up to the mid-1950s

In 1952, Ferrari wins the Mille Miglia once again. This time the 250 S is the winning sports car. It marks the beginning of the successful history of this model series

The model designations of the winning sports cars – 125, 166, 212, 250 and 340 – indicate the volume of a single combustion chamber. What they have in common is the distribution of the total displacement over a dozen cylinders. Gioacchino Colombo's engineers and technicians thus created a genetic fingerprint that is still firmly anchored in Ferrari's DNA today. Ferrari's relationship with Formula 1 is no less strong, with the Scuderia racing in the premier class from the very beginning in 1950. The first victories came in 1951, followed by the first titles in 1952, 1953 and 1956. World champions were Alberto Ascari and Juan Manuel Fangio (1956).

With the launch of the 250 GT in 1955, Ferrari laid the foundations for the small series production of road sports cars.

Mike Hawthorn drives his Ferrari 246 Dino F1 to second place at Silverstone in 1958. At the end of the season, the Briton wins the Formula 1 World Championship with Ferrari

At Le Mans in 1951, Ferrari privateer Jean Larivière was killed in an accident at the wheel of a 212 Export. The Frenchman is one of many fatalities that Ferrari will have to mourn over the decades

The Ferrari 125 S and 125 Spyder Corsa lined up side by side on the Circuito di Parma (located in the city) in 1947. Tazio Nuvolari wins the one-hour race on the three-kilometre track with the spartan-bodied racing car, one second ahead of his team-mate Franco Cortese

Ferrari has been one of the main players at the Targa Florio from the very beginning. The Scuderia wins four times with Alfa Romeo and seven times as a manufacturer

(1957 – 1966)

The second decade

Sport is Ferrari's elixir of life: for the company boss, participation in Formula 1 and sports car races is the purpose of the company. Series production served more to finance it. And the balance sheet is first class: In Formula 1, Ferrari is in the top tier. Mike Hawthorn, Phil Hill and John Surtees rushed to the world championship title with their 246 F1 (1958), 156 F1 (1961) and 158 (1964) respectively. With its first mid-engined monoposto, the Scuderia dominates the 1961 World Championship and secures the first of 16 titles in the Constructors' World Championship, which has been organised since 1958.

The Mille Miglia chapter closes in 1957 with a tragedy and another Ferrari victory.
Years of investigations into Enzo Ferrari following the devastating accident at the last event motivated Ferrari to transform Auto Costruzioni Ferrari into a public limited company in 1960. Its name: Società Esercizio Fabbriche Automobili e Corse, or SEFAC for short.

The team also collected a string of victories at Le Mans, and another four wins at the Targa Florio up to 1965. The winning sports cars were named 250 GT, Testa Rossa and 330, among others, and the 250 GTO was also among them. It becomes a winner and an icon. And a symbol of the end of an era, as successors such as the Dino 246 SP, 250 P and 275 P as well as the 250 LM were built with a mid-engine design.

With the 250 GTE, the programme now also included the first genuine production vehicle. Between 1960 and 1963, 955 of these were built.

In 1964 and 1965, Ferrari entered Formula 1 simultaneously with the eight- and twelve-cylinder 158 and 1512 models, with the 1.5-litre twelve-cylinder also sometimes referred to as the 512. The 312 followed in 1966

Ferrari homologates the 250 GTO as a further development of the 250 GT – without fulfilling the conditional obligations. This causes bad blood with the competition, but brings Ferrari numerous victories

The 250 P and 275 P marked the beginning of the era of twelve-cylinder mid-engined prototypes at Ferrari. They were also among the winners at Sebring, Le Mans and the Nürburgring

Rosso Corsa in front of Mare Azzurro: Wolfgang Graf Berghe von Trips drives a Ferrari 801 at the Monaco Grand Prix in 1957. He started here a total of six times in a Ferrari

The design of the 365 GTC (and its predecessor, the visually very similar 330 GTC) epitomises the further development of Pininfarina's Ferrari design language in the phase between classicism and modernity

(1967 – 1976)
The third decade

In Daytona, Ferrari celebrates a triple victory in the 24-hour race that attracts worldwide attention, and with the »Daytona« Ferrari wins customers like never before. In the first two decades up to 1968, a total of 6,000 vehicles are built in Maranello – as many as today in less than six months. With the Dino and the 365 GT/4 BB and 512 BB, the mid-engine era also begins on the road. And the trio of Niki Lauda, Mauro Forghieri and Luca di Montezemolo wrote a new chapter in Formula 1 history.

The design paths of racing cars and road cars gradually diverge. Nevertheless, the sports cars from Maranello celebrate pure racing technology – just domesticated for road use from the outset. This also applies to the Dino racing sports cars equipped with a V6 engine, from which two compact sister models for everyday use are derived: the 206 GT and the 246 GT, or GTS. With these more affordable mid-engined sports cars, Ferrari expanded its model range and tapped into a new clientele. A total of around 4000 examples of the small Ferrari are produced.

In 1972, the »Pista di Fiorano«, the test track opposite the factory, is put into operation. Enzo Ferrari moves into the building in the centre of the track and follows the development work of the F1 team, which is once again successfully reaching for world championship laurels.

In 1970, the armada of 512 S cars lined up in Maranello for homologation approval. Ferrari thus took its cue from Porsche and its 917, which was lined up for homologation in 1969

With the 308 GTB, Ferrari opened up new market potential from 1975. And the 365 GTB/4 »Daytona« Competizione is a promising sports car for private drivers

After a long dry spell, Niki Lauda finally brings Ferrari the world championship title again in 1975. 1976 also got off to a promising start. However, due to his serious accident at the Nürburgring, Lauda narrowly loses the title duel to McLaren driver James Hunt

(1977 – 1986)

The fourth decade

Niki Lauda and Jody Scheckter win another Formula 1 title. Gilles Villeneuve delights the Commendatore and the fans before subsequent Scuderia Ferrari drivers have to wait a long time for further laurels. The F1 team is working flat out to build on its past successes. In the back of the Ferrari 126 C2, or C3, with which the Scuderia won two more constructors' world championship titles in 1982 and 1983, a 1.5-litre six-cylinder engine with twin turbocharging was used. It produced 580 and 620 hp respectively. By the end of the 1988 season, the turbos were even pushing into the region of up to 1000 hp.

Turbo power was also used on the road: the 288 GTO presented in 1984 mobilised a breathtaking 400 horsepower and became the first production car in the world to exceed the 300 km/h mark. Its successor, the F40, was even 50 hp more powerful and accelerated to more than 320 km/h. It became a symbol of the worldwide hype that the brand experienced until the early 1990s.

The latest generation of the mid-engined Berlinetta, which also appeared in 1984, was also extreme. The flat 180° V12 in the Testarossa initially produced 390 hp, later up to 440 hp, and was concealed under a spectacular body characterised by horizontal lines and wedge-shaped rising side surfaces. As in Formula 1, the aerodynamics provide additional downforce. Thanks to ingenious wind tunnel work, the Testarossa can dispense with visually distracting spoilers.

Series production and Formula 1 tie up capacity in Maranello to such an extent that there is no room left for competition outside the premier class. It is therefore reserved for dedicated private drivers to win the Italian Rally Championship (1982), for example.

Gilles Villeneuve and Didier Pironi fought a relentless duel for supremacy in the team in 1982. In Imola, Villeneuve feels he has been denied victory. He then suffers a fatal accident in the following Belgian GP

With the Testarossa, in-house designer Pininfarina once again demonstrates automotive haute couture. It became one of the dream cars of the 1980s. The customised luggage set comes from Schedoni

Another dream car is the 288 GTO. At the time of its release, it was the world's fastest production sports car – and it founded the hypercar genre

(1987 – 1996)
The fifth decade

Just missed is still a loss: Ferrari has been waiting for the next Formula 1 champion since 1979

Harbingers of a unique era: Schumacher joins Ferrari in 1996

With successful drivers (Cesare Fiorio, John Barnard) and a four-time F1 champion (Alain Prost) on board, the Scuderia achieves further victories. Ferrari's 100th Formula 1 success is among the six victories of the 1990 season, but the team once again narrowly misses out on the title despite pioneering innovations (electro-hydraulically controlled gearbox that can be shifted from the steering wheel).

The death of Enzo Ferrari on 14 August 1988 accelerated the hype surrounding the brand, causing prices for all products from Maranello to skyrocket. As a result, the global sports car and classic car market also reached an unprecedented level of appreciation. In the end, speculators gave way to fascinated customers. Luca di Montezemolo instinctively steers the fortunes of the company and sets the course for success. The new models are technically interesting and stylish: the 456 GT is a classic Gran Turismo in the best sense of the word. With the F355, the aerodynamically shaped underbody makes its way into series production sports car construction, and the performance potential reaches new heights with 380 hp. In 1996, the front-engined Berlinetta 550 Maranello once again drew on classic models such as the 275 GTB or the »Daytona« – with even better performance. In addition, there is the anniversary model F50, derived in a straight line from the F1 racing car. Together, they manifest the commercial success of the company.

A turning point in Formula 1 was the reduction in the number of cylinders in the drive units to ten from 1996. The new engines also brought new drivers: Michael Schumacher and Eddie Irvine. The signing of Schumacher results in sporting successes, for which team boss Jean Todt prepares the ground as early as 1993.

On 12 July 1987, Ferrari presents itself and its fans with the F40 for its 40th birthday – a vehicle that remains exceptional to this day

(1997 – 2006)

The sixth decade

With meticulousness, dedication and total commitment, Gestione Sportiva matures into a team that dominates the Formula 1 world. Ferrari wins the constructors' title six times in a row. Around Jean Todt, Ross Brawn and Michael Schumacher, a team is formed that wins 63 Grands Prix in the years between 1999 and 2004 alone. They also won two titles in the FIA GT Championship. 22 victories make the 550 Maranello the most successful car in this series.

The models from the Gran Turismo product range also proved to be winners. Despite repeated adjustments to production capacity, the company is barely able to meet the demand for road-going sports cars.

With the F355 F1, the metallic clacking of the gearstick lever, which was previously guided through an open metal gate, gives way to silent pulling movements on paddle shifters mounted behind the steering wheel. The gearshift technology adopted from Formula 1 immediately achieves an installation rate of 90 per cent in road vehicles. New sports cars, such as the Enzo, the 599 GTB Fiorano, the 612 Scaglietti and the F430, make the brand more attractive than ever. In addition, a new F1 wind tunnel expands the research capacities of the series vehicle developers. On the other hand, Ferrari is now also involving end customers in its motorsport development activities with the FXX and the Corse Clienti programme.

Ferrari's professional drivers for the top league are Michael Schumacher, Luca Badoer and Rubens Barrichello (from left to right). Jean Todt is the Ferrari mastermind; anything but a shadowy figure

For a long time, refuelling was part of the pit stop. Since 2010, this has been put to history for safety's sake

Two masters of their trade: Michael Schumacher (with balaclava) and Todt on the starting grid

With the 550 GTS Maranello, Ferrari won the GT category at Le Mans in 2003, the FIA GT Championship in 2003 and 2004, and the LMES in 2004 and 2005

(2007 – 2016)
The seventh decade

Kimi Räikkönen replaces Michael Schumacher, who retires from F1 for the time being. With the Finn, Ferrari once again wins both F1 titles in 2007. The following year, team-mate Felipe Massa was able to feel like a world champion, but only for a few seconds, before the F1 crown went to McLaren-Mercedes driver Lewis Hamilton at the chaotic end of the season in São Paulo. Ferrari at least retained the Constructors' World Championship title.

Sebastian Vettel joins Scuderia Ferrari in 2015 – with four F1 titles under his belt. The aim: to share success in the footsteps of his idol Michael Schumacher.

Ferrari is equipping itself for the automotive challenges of the future with innovative approaches. This includes the 599 GTB HY-KERS concept study, the first hybrid sports car for the road.

Among the eight-cylinder models, the 488 Italia follows in the footsteps of the F430. With the FF, Ferrari presents a »Shooting Brake« for the first time, with all-wheel drive. »FF« stands for »Ferrari Four« – four driven wheels, four seats. The successor to the 599 GTB Fiorano is called the F12. As with the 599, various limited edition special models are also produced, which are very popular among collectors. At the lower end of the model range is the California, a convertible with a folding hardtop. The Ferrari LaFerrari crowned the model range in 2013 – Ferrari's first series-produced hybrid sports car, available as both a coupé and an Aperta (open top).

In 2015, the parent company Fiat-Chrysler successfully floats the sports car manufacturer on the stock market. Initial market capitalisation: around 10 billion euros.

With the La Ferrari, Maranello punctuates yet another exclamation mark. Number five of the »Ferrari Big Five« relies on hybrid power and sets the course for an electrifying future

Kimi Räikkönen wins Ferrari's last F1 drivers' title for the time being in 2007. Felipe Massa narrowly misses out in 2008

The front of the F430 is modelled on the shape of the »Supersqualo« F1 car from 1961. Like its successful role model, the eight-cylinder sports car also became a benchmark in its field: as the 430 GTC or 430 GT2, it collected successes like no other vehicle in this category

Ferrari wins Le Mans with hybrid power in 2023 – and at the first attempt. After five decades of factory abstinence, Ferrari wins its comeback in the top category ahead of Toyota and Porsche

Another first: the Frikadelli team gives Ferrari its first victory in the legendary 24-hour race on the Nürburgring-Nordschleife in 2023 with the 296 GT3

A permanent fixture for gentleman drivers on the world's racetracks: private teams and drivers win 407 of 770 races and no fewer than 107 championships worldwide with the 488 GT3

(since 2017)

The eighth decade

Special series prove to be a successful business model for Ferrari. In addition to various special models, Ferrari establishes the so-called »Icona« series. The Monza SP1 (single-seater) and SP2 (two-seater) models – extreme open sports cars without a roof and windows with 810 hp – were the first of these, limited to 499 units. This is followed by the Daytona SP3, also based on the 812 Superfast (the F12 successor).

The new eight-cylinder SF90 model is launched to coincide with the 90th anniversary of the founding of Scuderia Ferrari. The Portofino takes the place of the California, followed by the Roma in 2019. With the 296 GTB, Ferrari launches a six-cylinder for the first time since the Dino, now as a hybrid.In the course of 2023 the company will deliver more hybrids than pure combustion engines for the first time. Ferrari is sold out for years: »The order books are at the highest level,« says Ferrari sales boss Benedetto Vigna. And with the Purosangue, Maranello is presenting its first four-door model, 20 years after the Porsche Cayenne and five years after the Lamborghini Urus; SUV, it seems, is no longer taboo in Maranello...

In Formula 1, Ferrari will contest its 1000th Grand Prix in Mugello in 2020. Year after year, the team managed to win, but the Scuderia had to wait for titles. But in 2023, Ferrari made a sensational comeback at Le Mans with the 499 SP Hybrid – exactly 50 years after the last start at the venue. Ferrari also wins the 24-hour race at the Nürburgring – a first in the 54-year history of this equally legendary event.

The first fully electric Ferrari is announced for 2025. In 2030, Ferrari management expects a model mix of 40 per cent pure electric cars and hybrids, with only 20 per cent pure combustion engines.

Caption: With over 500 victories, the 488 GT3 is the most successful Ferrari in sporting terms.

With the Purosangue, Ferrari presents its first four-door sports car – Ferrari doesn't really like the term SUV...

With the XX models designed exclusively for racetracks, the brand is opening up a new, highly lucrative business segment

Space for four people is now just as much a part of the Ferrari engineers' specifications as is a modern electronics architecture and infotainment hub.

The 812 Competizione presented in 2021 is the limited edition upgrade of the 812 Superfast; the most powerful series production V12 with 830 hp. The naturally aspirated engine alone is responsible for its power – without additional electric power. The Monza and Daytona SP models are also based on the 812 Superfast

Enzo Ferrari began his career behind the wheel of competition cars. He had a very special relationship with racing drivers throughout his life; he even took some of them to his heart. But ultimately, only victories and titles counted for him

PILOTI, CHE GENTE...

Enzo is ten years young when he watches Vincenzo Lancia, Felice Nazzaro and co. drive the Coppa Florio race on the almost 53-kilometre Circuito di Bologna (Bologna – Castelfranco Emilia – Sant' Agata Bolognese – San Giovanni in Persiceto – Bologna) together with his father and brother. From then on, the passion for racing burnt in him. As a young man, Ferrari found company and work in the automotive industry. The racing driver Ugo Sivocci became a close friend. With his support, Enzo joined CMN and then Alfa Romeo the following year. Through his work as a transfer driver and in the wake of his friend, Ferrari also became a racing driver. However, Enzo Ferrari realised early on that his talent and ability did not match that of the absolute top drivers he admired. Antonio Ascari also became a good friend. The two got to know each other in 1920. And Ascari's work as a car dealer and Alfa general agent for Lombardy became a role model for Enzo.

Enzo ended his driving career when his son Alfredo (»Dino«) was born in 1932. »I can't swear that I would ever have become a good driver. I had my doubts even then,« he remarks self-critically. »I knew that I had a serious flaw as a driver: I had respect for every car I drove. If you want to achieve spectacular results, you have to know how to abuse your car. Hard shift manoeuvres, driving faster than the engine can handle and reckless braking; all things that go against my feeling for a car,« explains Enzo Ferrari, who was concerned with »feeling with the car, becoming a part of it.« It pains Enzo Ferrari »to see them die – because cars always die in a race, even if they win. That was something my emotions couldn't bear.«

In the long list of Ferrari drivers, there are a number of drivers that Enzo holds in high esteem and a few that he even adores. The latter undoubtedly included Tazio Nuvolari. The two met as rivals in 1924. The Ferrari brand is often referred to as the »Stradivarius of automobiles«.

Chris Amon is victorious, but the New Zealander always just misses out on the F1 title

Wolfgang Graf Berghe von Trips has a fatal accident in 1961 before winning the title

Michael Schumacher comes to Maranello with two world championship titles in his luggage. Together with Ross Brawn and other colleagues, they form a »dream team« and win eleven world championship titles together

Gerhard Berger experiences two three-year periods of service at Scuderia Ferrari

Enzo Ferrari admires the exceptional talent of Tazio Nuvolari. The picture shows Nuvolari in a Ferrari 125

Born in Lodi in 1930, Eugenio Castellotti (pictured here before the 1955 Mille Miglia) is one of Scuderia's sports car aces

It is therefore fitting that the young Nuvolari is regarded as the »Paganini of racing drivers«. He joined Scuderia Ferrari in 1930 after Giuseppe Campari returned to the Alfa works team. Enzo Ferrari paid for this change with Alfa Romeo P2s reworked by Vittorio Jano.

The »Flying Mantuan«, previously successful in motorbike racing, only took up motor racing at the age of 32; »il Campionissimo«, as Nuvolari is known in Italy, celebrated his greatest successes between 1933 and 1935, at the ages of 41 and 43 respectively. And even after the war, at the age of 55, he was still a world-class driver. »Nuvolari was already racing at a time when cars still resembled long-legged, stubborn wild donkeys and were not the sleek tiger cats of today that seem to slide along the ground,« wrote Porsche racing driver and journalist Richard von Frankenberg in the 1960s. Nuvolari is regarded as the inventor of the powerslide, the drift over all four wheels. He thrilled the public and characterised the generations of drivers that followed him. His driving style is the opposite of the clean lines of Achille Varzi. Incidentally, Vittorio Jano was reluctant to have Nuvolari in the Alfa team – »a madman who only ruins the cars…«. Later (1935), he became the mediator thanks to whom Nuvolari returned to the Scuderia.

Alberto Ascari became the first Ferrari world champion. The son of Antonio Ascari, who had been killed in an accident in 1925, won the F1 titles in 1952 and 1953, to whom Enzo Ferrari felt a similar bond as his former close friend Antonio. The pain is correspondingly deep when Alberto also loses his life in 1955 during test drives with a private Ferrari in Monza. Dr Giuseppe »Nino« Farina also belonged to the circle of Ferrari drivers. He became the first F1 champion with Alfa Romeo in 1950. The son of Giovanni Farina, head of the coachbuilding company Stabilimenti Farina, earned his doctorate in economics. He earned his first merits in his parents' company before the war – not to be confused with his uncle Battista »Pinin« Farina's coachbuilding business. Nevertheless, Stabilimenti Farina had to close its doors in 1953, not least because neither Nino nor his brother were able to adequately take care of the company's affairs.

Farina crossed paths with Juan Manuel Fangio at Alfa Romeo. The Argentinian already had various world championship titles under his belt

John Surtees was one of Enzo Ferrari's favourites – until he left the team in 1964

Enzo Ferrari in conversation with Luigi Musso, Eugenio Castellotti and Peter Collins

Kimi Räikkönen drives in red for eight years and wins a title

Juan Manuel Fangio joins Ferrari in 1956 and crowns his year at the Scuderia with title number four. But the liaison between Fangio and Ferrari is no love affair

when he joined Ferrari in 1956. It was not a love match for either Ferrari or Fangio, but after the withdrawal of Mercedes-Benz and Lancia, Fangio lacked promising alternatives. And the D50, inherited from Lancia and designed by Jano, was the best car of the year. In fact, Fangio managed to win a fourth title in 1956. However, his younger team-mate Peter Collins also played his part: The Brit handed over his car to the Argentinian at the race in Monza after his car broke down with a broken steering lever – even though Collins himself still had a chance of winning the world championship at this point. But he would never achieve this goal: Two years later, he had a fatal accident in his Ferrari during the 1000-km race at the Nürburgring.

It is said that for Enzo Ferrari, the drivers were merely subordinate to the technology. Accordingly, they were not included in meetings at the time. »Ferrari was convinced that there were enough young men who were prepared to risk their lives, their reputation and their inherited fortune behind the wheel of a first-class racing car. As long as his cars were racing, he didn't have to worry about the supply of drivers,« summarises Brock Yates.

Peter Collins' closest friend is Michael »Mike« Hawthorn. The British driver was haunted by the Le Mans tragedy of 1955, which he played a key role in triggering, for the rest of his life. But in 1958, he became the next world champion in a car from Maranello. At this time, there were a large number of names driving Ferraris in Formula 1 and sports car races. Doug Nye describes everyday life at the factory: »To even visit the factory there, you either needed an appointment or good connections, and the stuffy waiting rooms there were legendarily referred to as cells.« Even millionaires and future world champions have to queue there for their Ferrari.

Among them are Olivier Gendebien and Eugenio Castellotti, who each bought half a dozen Ferraris before becoming a works driver and had to queue for hours outside the factory. Wolfgang Graf Berghe von Trips also belonged to this group. In October 1956, he wrote to Enzo Ferrari: »Dear Mr Ferrari, ... regarding the Grand Tourisme car, may I ask you to let me know whether I can get one, what the price is and how you intend to settle the account. Or do you think it would be better for me to drive a works car on some occasions? If I can buy the car, I would run it at all the German events. If I'm unable myself, I'd have Seidel or Rolff drive it...« Enzo Ferrari replied just under a month later: »Dear Count Wolfgang, ...as far as the Grand Tourisme car is concerned, I can tell you that I am planning a super-light vehicle that will be ready in January or February. I agree that until my car is delivered, you can use another vehicle for passenger and personnel transport

With his unleashed driving style, Gilles Villeneuve wins the hearts of Ferrari and fans alike. But he remains without a title

Particularly during the 1950s and 1960s, Ferrari was able to play the queue of racing drivers off against each other

and for competing in races in your country or in your class, wherever you wish. On 1 December, I will be holding a party for all drivers and employees and I hope you will be there as my guest...«

Due to the sheer number of drivers available, they can never rest assured. Ferrari firmly believes that this uncertainty makes them better drivers. He respects them and uses every means at his disposal to keep them at the top of their game. »Gendebien remembers a scene typical of Ferrari and its way of manipulating drivers: Gendebien wanted to drive Musso's faster car, thought he had qualified for it. But Ferrari told him to drive the 250 GT,« says Ferrari chronicler Brock Yates. »Gendebien leaves the office in frustration and meets Alfonso de Portago, who is on his way to see Enzo. Five minutes later, Portago comes out, walks up to Gendebien and says, ›Ferrari says you want to drive my sports car; but the bastard also said it wouldn't make any difference anyway. He says you'd beat me anyway, no matter what car I'm in.‹« Alfonso de Portago is considered a daredevil. »I don't like the Mille Miglia,« he tells US journalist Ken Purdy a few weeks before the race, »No matter how much you train, it's completely impossible to ever know these thousand miles of country road as well as the Italians. As Fangio said: ›If you're not completely unscrupulous, you can't race like a madman here anyway. At the Mille Miglia, there are hundreds of corners where even the smallest driving mistake can cost fifty people their lives.« In 1957, at the checkpoint in the Apennines, Enzo Ferrari poured fuel on the fire once again: Gendebien would beat him... De Portago decided not to have his ailing tyres changed and hurriedly set off on the final spurt over the fast country roads of the Po Valley. At a good 290 kilometres an hour. Then came the accident that cost de Portago, his co-driver Nelson and nine spectators their lives. »No triumph or fame in the world is worth a single centimetre of human skin,« says Enzo Ferrari. But also: »Win or die, and you become immortal.«

Ferrari won its next Formula 1 title in 1961. However, Phil Hill becomes world champion, not his team-mate Wolfgang von Trips, who is killed in an accident while leading

Ferrari feels just as attached to Alberto Ascari as his father Antonio. Alberto wins Ferrari's first F1 title

A new era of professionalism begins in Maranello with Niki Lauda. Mario Andretti starts several times in red

Here Paolo Marzotto and Piero Taruffi talk to Enzo Ferrari in the run-up to the 1955 Mille Miglia

Eugenio Castellotti and Peter Collins (right) with a mechanic preparing the sports car for the race

the standings at the finale in Monza. Richard von Frankenberg describes the American as the »son of a postmaster from Santa Monica in California, who studied music before turning his hobby for fast cars into a profession.« In the team's internal race, he falls behind the carefree and fun-loving German.

Ferrari always managed to sign the best of the best. However, as with Fangio, the relationship with Stirling Moss was also strained. In Fangio's case, it was a failed car handover in 1949; in Stirling Moss' case, it was a Formula 2 commitment that was not honoured two years later. Moss only drove 13 races in a private Ferrari entered by whiskey millionaire Rob Walker.

John Surtees is the next F1 champion in red. Enzo Ferrari appreciates the motorbike world champion. Accordingly, the Briton is also closely associated with the family. Surtees returned the favour by winning the title in 1964. Despite his closeness to Laura and Enzo Ferrari, some of the customs in Maranello made Surtees shake his head: »Every salary negotiation with Enzo was difficult and protracted. When you finally got up the nerve to complain about the salary level, you always heard the same thing about ›honour‹.« Ultimately, differences with the then race director Eugenio Dragoni led to his departure before the end of the 1966 season.

A long dry spell followed for Ferrari in the premier class. Jacky Ickx was one of the drivers who ensured victories and title hopes in the early 1970s. However, the driver and team narrowly missed out on the world championship several times. »We need winners. We don't need second or third place finishers,« says Enzo Ferrari. A total of eleven years passed before the next title was won. In 1975, the Austrian Niki Lauda finally achieved the longed-for success. The following year, the fire accident at the Nürburgring and the rain chaos at the finale in Japan thwarted the title defence. Lauda and Ferrari triumph again in 1977. However, Niki does not forgive his »opponent« for the fact that Enzo Ferrari favours Carlos Reutemann over the champion, who is scarred by the accident, in the second half of the season. »I told him,« Lauda says, »that I had promised to drive for him for the rest of his life. But my feelings had changed and I asked him to release me. I no longer want to drive for you.«

Lauda left the team at the end of 1977 and took the starting number 1 with him to Brabham. Ferrari is furious. A week after the separation, he said in an interview: »Lauda is worse than Judas. He sells himself to the competition for 30 salami sticks.« Niki Lauda is quoted worldwide as saying that he is fed up with driving in circles. In fact, the then 27-year-old was extremely annoyed at suddenly being at a disadvantage compared to his replacement Carlos Reutemann. Decades later, Lauda admits: »That was the biggest mistake.« His compatriot Gerhard Berger explains why Lauda was so important and right for Ferrari at the time: »Niki Lauda was such an exceptional figure in the seventies: Because he did not allow himself to be bullied by Ferrari politics, but instead gained enough influence as a driver to actually utilise the superior technical resources. The company's technology always depended on the company's policy. If you wanted to make a difference as a driver, you had to be in the thick of the intrigue.«

Gilles Villeneuve replaces Lauda. A driver for whom Enzo feels a passion, as he once did for Nuvolari. The Canadian sets off fireworks and captivates the fans with his unleashed driving style. In the team's internal World Championship duel in 1979, however, he was defeated by South African Jody Scheckter, who took the next World Championship crown to Maranello. Villeneuve died in 1982 in Zolder, Belgium. The accident was preceded by an altercation with his team-mate at the time, Didier Pironi; the Canadian felt he had been robbed of victory in the Imola race. As Pironi also suffered a serious accident during the course of the year, the driver squad that season consisted of a total of four drivers, including Patrick Tambay and Mario Andretti. It was rumoured that the talented German youngster Stefan Bellof had already signed a contract for the 1986 season. But his hopes were dashed by his Porsche accident at Spa-Francorchamps in 1985. At that time, Gerhard Berger joined Ferrari: »In any case, I was the last racing driver that Enzo Ferrari personally selected.«

Long-time McLaren designer Steve Nichols experienced the time in Maranello after the death of Enzo Ferrari in 1988. For him, Ferrari – compared to the British teams – was more or less in a state of constant chaos. »That didn't make it easy for the drivers. Prost spent a year there. He is calculating, cool, calm and very methodical. Alesi is southern, emotional, and that's not necessarily the best thing in a team like that.« But Jean Alesi stays with the Scuderia for five seasons. Alain Prost, on the other hand, was shown the door in 1990. His mistake: he tried to play team politics via the media. Prost says that »internal crises at Ferrari are completely normal.« And criticised both his team boss at the time, Cesare Fiorio, and the car. The retort followed immediately. Also via the media. Umberto Agnelli, the brother of Fiat boss Gianni, spoke out: »Senna is the best driver in the field, and that's why I would like to have him at Ferrari. Of course Prost is a great driver, but he is going through a difficult time and seems to have become less aggressive. Also, his enthusiasm leaves a lot to be desired.«

Behind the scenes, the way is clear for the formation of Montezemolo, Todt, Brawn and Schumacher. And with Michael Schumacher's move to Maranello, Berger and Alesi take their hats off and join Benetton. Eddie Irvine, Rubens Barrichello and Felipe Massa drive alongside »Schumi« in the coming years, which are characterised by the Kerpen native. The Schumacher years yielded six constructors' titles and five drivers' titles. The record-breaking champion leaves Ferrari at the end of the 2006 season, with the Finn Kimi Räikkönen winning another title in red the following year. Felipe Massa (2006-2013), Fernando Alonso (2010-2014), Sebastian Vettel (2015-2020), Charles Leclerc (since 2019) and Carlos Sainz junior (since 2021) did not achieve this. Enzo Ferrari would probably comment on this as follows: »Bad luck is the consequence of what we did not achieve or could not foresee."

Racing driver and importer: Luigi Chinetti revitalises Ferrari sales in the New World with his racing successes

Jubilation in red: Eddie Irvine (centre) is Michael Schumacher's first team-mate. The Irishman is followed by Rubens Barrichello and Felipe Massa

The Barchetta, with its body designed by Carrozzeria Touring, became Ferrari's success story of the early years. The majority of the first models, however, were fitted with bodies from Vignale - such as the 225 Sport Vignale

THE FIRST FACE IN THE CROWD

The increase in displacement of the twelve-cylinder engine, designed by Gioacchino Colombo, to 2 litres – 166 cm^3 per combustion chamber – marked the historic starting point of the model that made Ferrari world-famous, with numerous racing successes within a very short period of time. It began with victories in the Targa Florio and Mille Miglia in 1948. Incidentally, from 1948 to 1950 the »Targa« was held on the »Giro di Sicilia« route, which covered the entire island – as it had been between 1912 and 1914, but now in an anti-clockwise direction. The open Ferrari 166 S of Clemente Biondetti and Igor Troubetzkoy, alias »Principe Igor«, coachbuilt by Allemano, crossed the finish line of the first post-war event as the winner after 1080 kilometres and twelve hours and ten minutes of driving time. The second 166 S (Besana/Sterzi) finished in sixth place. Four weeks later, at the beginning of May, Biondetti and his co-driver Navone then won in the 166 S Allemano Coupé. The former Alfa Romeo driver took Ferrari's first »Mille« victory. However, all eyes were on Tazio Nuvolari. The »flying Mantuan« also drives a Ferrari. Enzo Ferrari visits the legendary driver, who is now seriously ill, and is able to convince him to start in one of his cars. Nuvolari is given a 166 SC to do so, but after a furious drive, he drops out of the race while in the lead, exhausted and with a bent chassis. It was to be his last Mille Miglia, and the long-awaited victory slipped through Nuvolari's fingers.

The following year saw two more successes for the Ferrari 166 at the »Targa« and »Mille«. Both times, the winner was once again Clemente Biondetti – in Sicily together with Aldo Benedetti, and in the Mille Miglia once again with Giuseppe Navone. At the Targa Florio, Biondetti again took the wheel of a minimalist Allemano Spider. For the Mille Miglia, Carrozzeria Touring then fitted a Spider with a striking Barchetta body; its designation was 166 MM. 28 minutes after the winning car, a second 166 MM (with the driver pairing Bonetto/Carpani) reached the finish line in Brescia. Among Ferraristi, this beautiful car is simply known as the »Barchetta«. The victory at Le Mans in 1949 also stands out in the CV of this model. Luigi Chinetti contested the first post-war edition of the 24-hour classic almost single-handedly – his co-driver, »Lord Selsdon,« dropped out after less than an hour of driving due to illness. As far as the Ferrari 166's track record is concerned, almost five dozen overall victories and a large number of class victories in prominent events were achieved by the mid-1950s.

In this early era, every single car that left Maranello was unique. In addition to racing

The 225 Sport produced 14 Spiders with Vignale bodywork. Here you can see the vehicle shown in the Loh Collection at the 1952 Mille Miglia. The Marzotto/Marini duo are in the cockpit. Their hurried journey ends 70 kilometres before the finish line

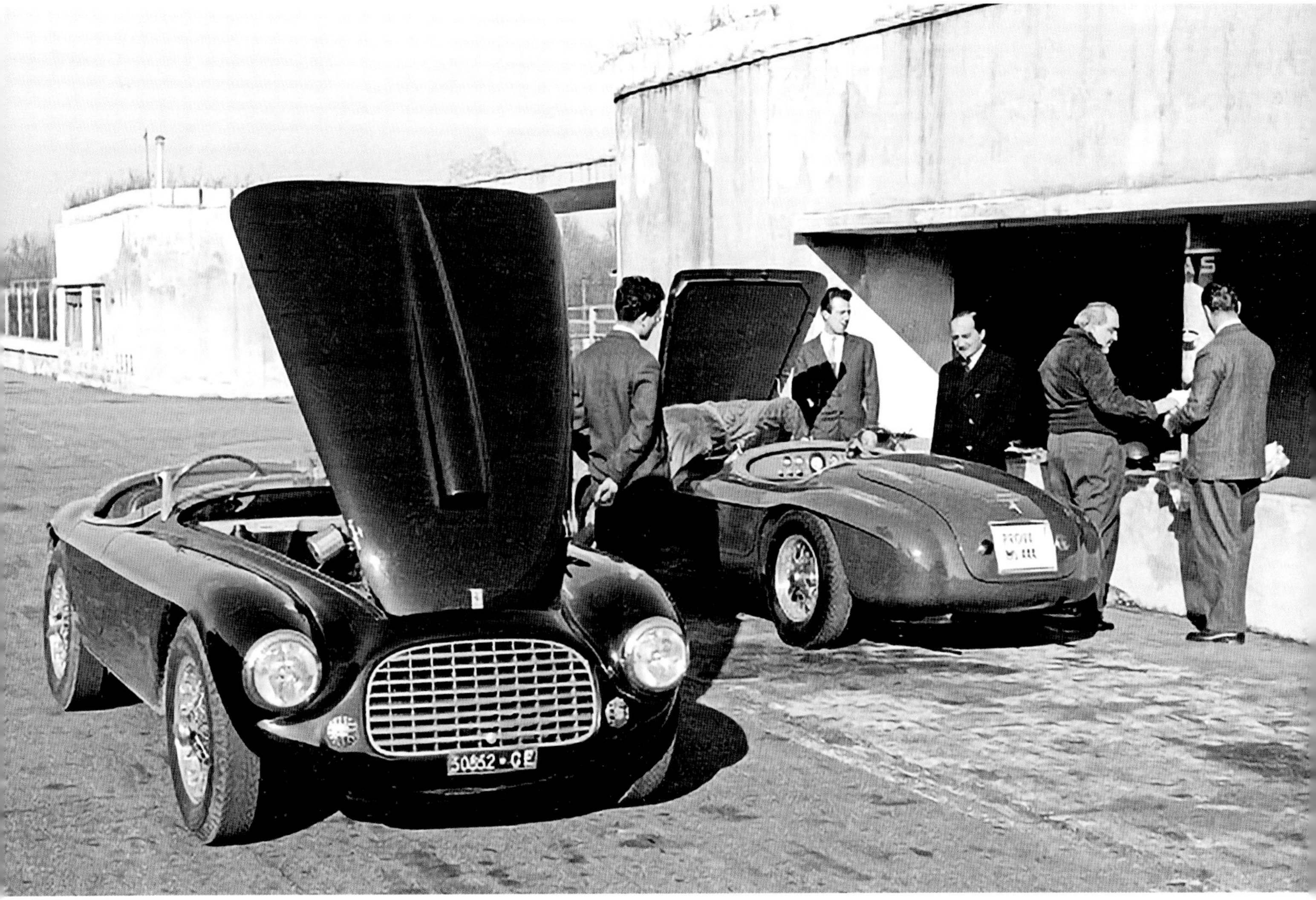

The »Barchetta« dressed by Carrozzeria Touring makes a striking impression. The face of the 166 MM and its sister model with coupé bodywork and larger displacement gives the brand its first face

cars, the 166 also produced its first road cars. Depending on the version, their engine output was between 90 and 160 hp. Their bodies also differed. The spectrum ranges from the 2+2-seater coupé to the tightly cut Spider in Superleggera dress from the touring workshop. Touring patented this design with a thin aluminium skin applied to a filigree tubular frame in 1937.

But Allemano, Bertone, Stabilimenti Farina, Ghia, Pinin Farina, Vignale and Zagato also produced various 166s with open and closed bodies. However, this model series was characterised by the appearance created by the touring designers led by Carlo Felice Bianchi Anderloni. In addition to the open two-seater, there was also a Berlinetta variant and an even more spacious coupé with space for two plus two people. Looking at the elegant face of the Touring 166, the head of the company says: »This required enthusiasm and courage: enthusiasm, because the aim was to give the Ferrari its own, unmistakable face«, and »courage, because the result was to revolutionise the conventional rules of sports car construction.«

In addition to the models that were successful in racing, there were also luxuriously equipped models, known as »Barchetta Lusso«. They are characterised first and foremost by an elegantly finished leather interior. In addition to a black trade fair model with a green interior and beige contrasting stitching, a touring Barchetta painted in deep blue and green also achieved prominence: its owner was the young Fiat heir Giovanni »Gianni« Agnelli. The colour scheme was what the playboy wanted at the time, as it reminded him of the colour of the sea off Capri. And so this model marked the beginning of great sales success for Ferrari in high-society and jet-set circles.

The successful Touring design was also used in the more powerful 212 MM and 340 America models. With the 170 hp 212 Barchetta, Phil Hill, for example, contested his first car race with a Ferrari in San Francisco's Golden Gate Park at the end of May 1952.
Another open Ferrari 225 contributes to Conte Paolo Marzotto becoming Italian champion

in 1952: The 225 Sport Vignale with chassis number #0172ET shown in the Loh Collection.

At the first start of the famous Mille Miglia, the gentleman driver from the textile industry – in third place – dropped out 70 kilometers before the finish. However, three victories in a row followed before the vehicle enjoyed its second career until 1967 in Argentina. The body of the open-top two-seater designed by Vignale was far simpler than the more advanced Touring design. But at the time, the well-known coachbuilder from Turin was doing good business in Maranello.

The 225 S comes in the darker red colour characteristic of Scuderia Ferrari. The 225 Sport Vignale is one of the most successful Spiders from Maranello. It wins a series of races; and its driver wins the Italian championship with it

AUTO AVIO COSTRUZIONI 815

After his separation from Alfa Romeo, Enzo Ferrari had to accept a four-year non-competition clause in return for the proverbial »golden handshake«. But when Alberto Ascari and his friend Marchese Lotario Rangoni Machiavelli asked him to build a car for the Mille Miglia in December 1939, Ferrari could not be stopped. Together with loyal companions such as Alberto Massimino, Ferrari's most experienced designer, and Enrico Nardi, a new sports car was created in record time.

In order to circumvent the non-compete clause, he founded the company Auto Avio Costruzioni. For the car, its builders drew on the parts pool of the Fiat 508C Balilla. A 1.5-litre in-line eight-cylinder engine designed by Massimino and a Superleggera body were also used. Ferrari chose the number combination 815 for the designation, and the model also appeared in the newspapers as the »Torpedo Brescia«.

In April 1940, two 815s lined up at the start of the Mille Miglia, which was exceptionally held on the 165-kilometre triangular course Brescia – Cremona – Mantua – Brescia. Ten laps make up the classic distance. The two cars will be driven by Alberto Ascari and his cousin Giovanni Minozzi as well as Rangoni/Nardi in the race, also known as the »Gran Premio di Brescia«. And they got off to a promising start: Ascari led the first lap with the new design. The sister car follows close behind and takes the lead after Ascari's retirement. But the second 815 does not make the finish line either. Nevertheless, the new design impressively demonstrates the potential of a sports car designed by Enzo Ferrari prior to going public.

Two months later, Italy is also at war. The lights go out in Europe and Ferrari produces machine tools. But he was already dreaming of a new beginning...

FERRARI 125 S

»The first Ferrari,« recalled Enzo Ferrari, »was a conventional car. There were no experiments. We wanted to build nothing more than a conventional engine – but it had to be outstanding.« It appeared in the spring of 1947 and established the history of the Ferrari brand as an automobile manufacturer.

In 1943, in the middle of the war, Enzo Ferrari moved from Modena to nearby Maranello. Reconstruction of the damaged factory began there in 1946. And he soon began a new chapter: Now the focus is on designing and building cars – but under its own name. However, even before the project took shape, Ferrari lost its long-time companion Alberto Massimino to Maserati. But he quickly found a replacement in Gioacchino Colombo. Ferrari also shared a common past with him; Colombo had been a member of Jano's development team for a long time and became Alfa's liaison to Scuderia Ferrari from 1933. He was soon joined by former employees Giuseppe Busso and Luigi Bazzi.

With a displacement of one and a half litres, the engine had the same combustion chamber volume as the Alfetta and the Auto Avio Costruzioni 815, but Ferrari and Colombo knew from the outset that this time it had to be a 12-cylinder. The Alfa Romeo S10, developed in 1939, provided a model for this. With its 60° V12, it has clear parallels to the Ferrari engine. However, the Ferrari V12, made of aluminium, is designed with a short stroke – and therefore high revving.

Power is transmitted via a 5-speed gearbox. The chassis design is based on Colombo's Alfetta calculations. And at 650 kilos, the small 1.5-litre sports car was a lightweight. Ferrari, on the other hand, thought big right from the start: »Ever since I built the first 1500, I had the ambition to start series production of high-quality cars.«

FERRARI 340 AMERICA »BARCHETTA«

The small »Barchetta«, a spider with touring bodywork, celebrated its debut in 1948 with a bang thanks to successes at the Targa Florio, Mille Miglia and Le Mans. Under the bonnet of the 166 MM is the 2-litre V12 from the pen of Gioacchino Colombo. At the same time, Ferrari commissioned engineer Aurelio Lampredi to design further twelve-cylinder engines. Background: Enzo Ferrari was determined to beat Alfa Romeo in Formula 1. The result was a 335 hp 4.1-litre engine, which made its first appearance in Ferrari's Formula 1 racing car 340 F1 in 1950.

As there was already such an attractive engine, it made sense for Ferrari to offer a matching sports car. In the late summer of 1950, the first brochures of the 340 America, a two-seater with a Lampredi engine domesticated for road use, were printed. In October of that year, an example was on display at the Paris Motor Show. At first glance, it looks very familiar: the trade fair model comes as a Barchetta with the familiar Touring body in Superleggera design. However, at 2420 millimetres, the 340 America has the longer wheelbase of the 166 Inter models; whereas the wheelbase of the 166 Barchetta, on the other hand, measures 2.2 metres. The vehicle – which is also produced with additional bodies – weighs just 900 kilos. This gives the 230 hp of the »big« Lampredi engine an easy ride: the 60° V12 enables a breathtaking top speed of 240 km/h. The smaller Barchetta, the 166 MM, weighs 100 kilos less, but with 140 hp it »only« reaches 185 km/h. In both cases, by the way, the gearbox is a 5-speed gearbox and the brakes are hydraulically operated drum brakes. With the 340 America, Ferrari – as the name suggests – is aiming straight for the seemingly tempting US market.

Ferrari has always relied on motorsport as a business model, test laboratory and sales-promoting opportunity. Hardly any stage is prominent enough for Ferrari. Success at the Carrera Panamericana boosts sales in the New World

CONQUEROR OF A NEW WORLD

Luigi Chinetti won the 24 Hours of Le Mans twice in 1932 and 1934. At the time, he was 31 and 33 years old respectively and was at the wheel of an Alfa Romeo 8C 2300. Luigi's father, a patented mechanic, achieved middle-class prosperity. Chinetti grew up near Milan. He enjoyed spending his afternoons at school in his father's workshop. After the First World War, Luigi joined Alfa Romeo. There he worked in the team of designer Giuseppe Merosi and also got to know Enzo Ferrari, among others. In 1925, Chinetti became part of the Alfa racing department. When he was allowed to accompany the racing team to Montlhéry for the French GP, he decided to learn French. He felt so comfortable there that he eventually moved to France and worked as a car salesman for the Alfa agency there. This connection also brought him to Le Mans. In France, the northern Italian also met the wealthy American team owner Laurie Schell in 1939. He travelled to the States for him in 1940 to support his team in Indianapolis and other races. When Chinetti wanted to return home, his path was blocked by the war. Chinetti was forced to stay overseas. He did not return to Paris until 1945, but found nothing there that dated back to his pre-war days. And so he visits his former friend Enzo Ferrari.

The success story in America begins with the double victory at the Carrera Panamericana

In Maranello, Chinetti and Ferrari talk about the Avio Costruzioni 815 car and the idle machine tool production. The situation is depressing, but Chinetti makes a suggestion to his friend: »I have good connections in France and America. You should come with your own car.« Enzo Ferrari replied that Chinetti would have to order at least five cars from him. »Five? I'll sell more than 20 of them,« Chinetti replies. The rest is history. And Chinetti becomes a part of it. Of the 14 authorised Ferrari dealerships at the beginning of the 1950s, Chinetti alone owned two (in Paris and New York).

On top of this, he founded the North American Racing Team (NART) in 1957 with the support of two financial backers. Its successes further publicised the Italian racers. Luigi Chinetti – who won Le Mans for the third time in 1949, now in a Ferrari – infected the Americans with the red virus. He was supported by enthusiastic »gentleman drivers« such as Jim Kimberley, Temple Buell, John Edgar and Phil Hill. The latter even made it to the F1 crown with the red cars in 1961...

»Win on Sunday, sell on Monday«: As the number of successes increases, so does the number of cars sold. Hill's successes are just as much a sales booster as Chinetti's Ferrari victory in the Carrera Panamericana ten years earlier. Together with his co-driver Piero Taruffi, Chinetti achieved the first internationally significant Ferrari victory on American soil.

However, participation in the race was to be as a privateer, as Enzo Ferrari was »not the least bit enthusiastic« about taking part in the Carrera, Chinetti recalls. Initially, only touring cars were allowed to take part in the race, which lasted a week and covered more than 3000 kilometres. At least four seats were mandatory. The two Ferrari 212 Inter cars entered in 1951 fulfil this requirement. Next to the huge road cruisers from US production, however, the petite Italian sports cars looked like undersized toys.

Unlike the 212 Export, the Inter models (for International) were originally designed as road-going sports cars. They have a 35 centimetre longer wheelbase and mostly Ghia or Vignale bodies. Giovanni Michelotti, who was not even 30 years old and worked there as a freelancer, was responsible for the lines of the Vignale bodies.

The designation 212 refers to the displacement of the 2.6-litre V12. In Carrera trim, it mobilises around 165 hp. On the rough asphalt and in

Like a time capsule — while the sister car has been considered lost since a serious accident, the carefully preserved and carefully restored #161 can be seen today in a fascinating condition

In the New World, advertising lettering has long been part of motorsport. In Europe, these barriers only fell after 1968

the Mexican heat, this is a challenge that Ferrari's preferred Pirelli tyres are not up to. The problem is only solved by changing to tyres actually intended for trucks – fortunately early on in the race. In order to be able to keep an eye on the tyres while on the road, holes are cut into the wheel arches. The driver and co-driver can see the rubber through the peepholes from their seats. In the end, Chinetti/Taruffi won ahead of their team-mates, the then F1 vice-champion Alberto Ascari and Mille Miglia winner Luigi Villoresi.

Unlike in Europe, advertising was already permitted in the competitive environment on the other side of the Atlantic; this was not approved in European organised racing until 1968. In the Carrera, the US oil company Sinclair made a significant contribution to financing the Ferrari adventure on the mostly murderous Mexican tracks. The race is also extremely well endowed: The victory bonuses and the proceeds from the race cars sold at the event will add 70,000 US dollars to the team's coffers. This makes the outing a thoroughly rewarding affair.

What looks like child's play at first glance turns out to be quite chaotic in practice: the two 212 Inter reach their ferry at the last minute. To prevent this from happening again the following year, Chinetti personally travelled to Maranello in the run-up to the 1952 Carrera to take delivery of the successor models to the previous year's winners – two 340 Mexico. There, however, Chinetti and Ferrari first came to blows over the terms of payment. Ferrari insists on immediate payment of the 2,200 US dollars due. When the money arrived in Maranello, it was no less high time for the new racers to embark than in 1951.

Incidentally, the previous year's racer changed hands in May 1952 for 2600 dollars, and the new owner brought the 212 to the start of the Carrera in 1952. In the end, the Mexicans Paco Ibarra/Vincente Solar achieved seventh place in the class and 24th place overall with the car, which is now part of the Loh Collection.

The quality perception of Ferrari customers at the time was certainly put to the test: »Even when the first Ferraris were produced in small series, the individual vehicles were anything but identical, neither in terms of styling, and often not even in terms of their dimensions,« explains Ferrari chronicler Pete Lyons. »In the age of hand-built bodies, symmetry was a foreign concept. For example, a right-hand wing was not necessarily the mirror image of the left-hand one, and wheel cut-outs and door openings sometimes varied by several centimetres. Ferrari owners

The rear section of the coupé produced by Vignale looks bizarre. At the time, the young designer Giovanni Michelotti was working in Alfredo Vignale's Turin workshop

who have ever seen their favourite car without any paintwork were in tears at the countless welded-in pieces of sheet metal and the amount of filler used.« And yet the wealthy playboys and wealthy gentlemen drivers of high society queued up at Chinetti in the 1960s. Chequebook in hand, they asked for the generous allocation of a Ferrari.

In Germany, Wilhelm alias »Auto« Becker is the Ferrari importer of the early days. In 1956, he reached an agreement with Enzo Ferrari for exclusive distribution in the Federal Republic. The contract is preceded by a brief exchange of letters that ends with the following lines: »We are pleased to inform you that our company, in response to your request, is willing to conclude a contract with your company for the importation and distribution of our production.« Auto Becker is already an importer of brands such as Renault, Alfa Romeo and Lancia. »Business relations with Ferrari developed extremely successfully,« says Wilhelm Becker. »A personal friendship grew out of this – despite the language barriers.« Ferrari later divided up the German sales territory and, since 1988 (when Ferrari Deutschland GmbH was founded), has relied on its own in-house sales network. This development also took place in the US market, where Ferrari has maintained contact with its customers through its own representative since 1981. In 2023, Ferrari will be selling its sports cars worldwide via 177 authorised dealers in more than 60 markets.

Occupants can check the condition of the tyres while driving through peepholes in the wheel arches

The headlights are a little too delicately dimensioned for use in Central America

With around 165 hp, the V12 is weaker than many US V8s. Ultimately, it will supersede them

Two full-size instrument clusters and the three-spoke steering wheel with wooden rim characterise the dashboard of the 212 Inter

For Enzo Ferrari, the »Mille« is an emotional affair, »because large parts of my life are linked to it. It ushered in a new era and shaped our cars. Without it, we would not have been able to develop our GT cars.«

A THOUSAND MILES OF EMOTIONS

For Enzo Ferrari, the Mille Miglia is an emotional affair, »because large parts of my life are linked to it«, as the entrepreneur summarises. »It ushered in a new era and shaped our cars and Italian motorsport. Without the Mille Miglia, we would not have been able to develop our GT vehicles, which we now sell worldwide.«

In December 1926, an endurance race on Italy's roads was discussed publicly for the first time when it was first written in Italy's important daily sports newspaper: the »Gazzetta dello Sport«. The considerations come at the right time. And so Mr Maggi, Mr Mazzatti, Mr Canestrini and Mr Castagneto were able to set the course for the race that would become the queen of all road races between 1927 and 1957. It runs in a loop from Brescia via Rome to Brescia; over 1000 miles – Mille Miglia. The starting fields were huge, as was the range of vehicles taking part. The smallest vehicles include cars such as the Fiat 500 »Topolino« and its racing offshoots.

Luigi Musso in the 860 Monza at the technical inspection before the start in 1956

Preparations for the start of the 1956 »Mille«, including Fangio's 290 MM (600), Collins' 860 Monza (551), a 250 GT (453) and 250 TdF (448)

Tragic end to a business trip – the wrecked car of de Portago/Nelson after the terrible accident with a total of eleven fatalities

The top dogs include GTs and sports cars from Alfa Romeo, Ferrari and Maserati.
The best time for the 1600 kilometre race distance was set by the Stirling Moss/Denis Jenkinson team. They raced through Italy in 1955 at an average speed of almost 158 kilometres per hour. After just 10 hours and 7 minutes, they are back in Brescia in their Mercedes-Benz 300 SLR. The Alfa duo Biondetti/Stefani had already beaten the 12-hour mark in 1938. It is no coincidence that the race is feared by the drivers. In the eyes of Jacky Ickx, »life is condensed into a few hours«, as the teams spend half a day on the most demanding, largely public roads, relentlessly going toe-to-toe with death. Enzo Ferrari is of the opinion: »No driver has ever really earned the laurels of victory unless he has won in Brescia.« Whilst Enzo Ferrari himself took part in the Targa Florio five times, he did not start once in the Mille Miglia.

When the idea was born, Italy was still an aristocratic monarchy. But Benito Mussolini was already beginning to make inroads into the political arena. In 1926, for example, he had the Azienda Generali Italiani dei Petroli (AGIP) founded – to supply fuel for the 170,000 cars on Italy's roads (in Germany, there were a quarter of a million at the time). Turin and Milan were also the last cities to switch to right-hand traffic in 1926 – prior to this, traffic on Italy's country roads and in smaller towns had been on the right, whereas in larger cities it had been on the left. Since 1922, Enzo Ferrari has earned his living as Alfa Romeo general agent for Emilia, Romagna and Marche. And he still takes the wheel of the cars from Portello from time to time as a racing driver. In the early years of the »Coppa Mille Miglia«, however, Enzo Ferrari organised service points for the Alfa drivers. Initially at the Raticosa Pass, a challenging mountain route on the border between Tuscany and Emilia-Romagna, and later also in San Casciano in the Val di Pesa. In December 1929, Scuderia Ferrari was founded. The company's objective is »the purchase of Alfa Romeo racing cars and their participation in races«. The 1930 Mille Miglia was the debut of the racing team,

This 250 GT goes by the name of TdF. But it is also a good choice for the Mille Miglia. Olivier Gendebien finishes a sensational third in 1957

Photographer Louis Klemantaski sits alongside Peter Collins in '56 and '57. Here they immediately trail the leading Count Berghe von Trips

In 1957, the Scuderia mechanics accompanied Wolfgang von Trips' Ferrari 315 S to the start in Brescia, Italy, which was set for 5.32 a.m. according to the starting number

Giovanni Bracco and Umberto Maglioli bring their 166 MM Barchetta back to Brescia in fourth place in 1950, clearly marked by the exertions of the 1000 miles

The colour photographs from the moving racing car were a sensation at the time. They impressively convey the speed; on long straights even beyond the 300 km/h mark

which quickly became famous. Between 1933 and 1937, the Scuderia Ferrari teams won the »Mille« five times in succession.

In the following years, Enzo Ferrari was responsible for the fortunes of the Alfa works team Alfa Corse. These »Mille« editions were also organised by Alfa Romeo and Ferrari. At the end of 1939, after leaving Alfa, Enzo Ferrari was asked by Alberto Ascari whether he could build two cars for the following year's Mille Miglia. No sooner said than done: two examples of the sports car known as the Auto Avio Costruzioni 815 were built – still incognito, due to the four-year non-competition clause agreed with Alfa Romeo. But in the hands of Ascari and Rangoni, these vehicles were already attracting the attention of experts in 1940.

At the first post-war Mille Miglia, a Ferrari also bearing this name was on the starting grid. While the 125 S driven by Cortese suffered a retirement on its debut, the following year – Clemente Biondetti in a closed 166 S – recorded a victory, the first of a total of eight »Mille« successes. Ferrari also became the brand to beat on this terrain – only Lancia and Mercedes-Benz took the crown in 1954 and 1955.

The history of this road race came to an abrupt end in 1957 when Ferrari driver Alfonso de Portago left the road at almost 300 km/h in the northern Italian lowlands near Guidizzolo. Along with the driver, his co-driver and nine spectators also lost their lives. To this day, the opinion persists that Enzo Ferrari egged his driver on and thus drove him to his death. What is certain is that mind games are deeply rooted in Enzo Ferrari's everyday life and the company. And that Alfonso de Portago, who was fighting for victory with his team-mates Wolfgang von Trips and Piero Taruffi, decided not to change his worn front tyres at the last service stop out of haste.

Anecdote in passing: Wolfgang Graf Berghe von Trips generously gives way to his much older team-mate Piero Taruffi before the finish. »Then I'll just win next year,« said von Trips. But that was not to happen. The terrible accident has caused a stir in Italy. Even the Vatican spoke out and called for this outdated race to be banned. In addition to the media coverage, Enzo Ferrari was also charged by the public prosecutor's office. The trial only ends four years later – with an acquittal for Ferrari. However, the story of the Mille Miglia ends in 1957.

In 1977, just in time for the 50th anniversary of the legendary road race, the »Mille« experienced its renaissance as a highly regarded international classic event. Year after year, participants from all over the world compete for the coveted starting places. Only vehicle models that once took part in the original 1000 miles are eligible to start.

TARGA FLORIO

The Florio family owns extensive land in Sicily and a number of companies. These include a shipping company. And so it was only natural for Vincenzo Florio to link the one with the other: he donated a trophy for a car race held on his land, which would soon attract participants from all over – and thus also keep his Sicilian ferries busy. The history of the Targa Florio began in 1906, and the race soon became one of the most important competitions in motorsport. And while races on public roads in other parts of the world no longer fit in with the times (and the volume of traffic) as early as the mid-1950s, the »Targa« history as a sports car world championship race lasted until 1977, thanks to the remoteness of the Sicilian hinterland.

Between 1919 and 1923, Enzo Ferrari entered the race five times, initially in a CMN and then in an Alfa Romeo. Even the journey through Italy was spectacular and challenging. Ferrari vividly recalls: »On the Abruzzo plateau, we got caught in a heavy snowstorm. Chased by wolves, we had to face completely unforeseen dangers. The wolves were chased apart by the intervention of the road wardens, who chased them away with guns and torches, and I also eagerly fired shots from my revolver, which I always kept concealed under the seat cushion of my car.«

Ferrari misses out on victory as a driver. However, he won four times with Scuderia Ferrari (1932-1935) and seven times as a manufacturer (1948, 1949, 1958, 1961, 1962, 1965 and 1972). Incidentally, Enzo Ferrari's friend Ugo Sivocci adorned his victorious Alfa Romeo RL with the Quadrifoglio Verde, the four-leaf clover, for the first time at the Targa Florio in April 1923. The lucky charm became a symbol of Alfa Romeo that is still important today. Two months later, Ferrari also uses a mascot – the Cavallino Rampante.

CARRERA PANAMERICANA

It takes 14 years to build the ribbon of tarmac that connects Mexico with Texas in the north and Guatemala in the south. The new road becomes part of the Panamericana that runs through the Americas and represents 25 per cent of Mexico's road network in one fell swoop. Out of pride and to communicate the new infrastructure, the Mexicans organised a race on the new main road for the first time in 1950.

The Carrera Panamericana is not just a national event, it becomes one of the greatest adventures in motorsport. The race starts on Independence Day. And the race, divided into several daily stages, takes place over the course of a week. It covers a distance of almost 3,500 kilometres over challenging roads and passes that rise to over 3,000 metres above sea level. The route is initially travelled from north to south. Then, from 1951 onwards, in the opposite direction.

Initially, the Carrera Panamericana was advertised for touring cars with at least five seats, like those rolling off the assembly lines in Detroit. But as early as 1951, the large-calibre US road cruisers came up against sleek European GTs. From then on, they would retain the upper hand in this race. The competition became a flagship event, ideal for product advertising on both sides of the pond. In 1953 and 1954, the »Carrera« even became part of the World Sports Car Championship. But then it came to an end: racing on public roads at speeds of up to 300 km/h had long been far too dangerous. Porsche continues to capitalise on the event's nimbus to this day – from the »Carrera« to the »Panamera«. Ferrari remains the most successful brand on the Central American circuit. Piero Taruffi and Luigi Chinetti won in 1951 with a 212 Inter, and Umberto Maglioli in 1954 with a 375 Plus Pinin Farina. The Lancia D24, designed by Vittorio Jano, wins in 1953 (Fangio/Bronzoni).

TOUR DE FRANCE (TOUR AUTO)

When most people think of the Tour de France (TdF), they think of the world-famous cycling race that takes place over three weeks and around 3500 kilometres through France. In fact, the automobile race of the same name can look back on a history that is four years older: While the cycle race was launched in 1903, the Automobile Club de France, founded in 1895, organised a race for the first time in 1899.

The first iteration covered 2200 kilometres, from Paris via Vichy and Nantes back to the capital. Over the years, the route grew to over 5000 kilometres. Due to the increasing traffic density on public roads, race tracks are also integrated into the event. The automotive Tour de France thus evolved from a road race to a mixture of mountain and circuit races, including regularity tests, to a rally. The great era of the »Tour Auto« began in 1951 and lasted until the mid-1960s. Ferrari dominated the initial post-war years: three 212 Exports took first to third place. In 1956, the Ferrari 250 celebrated its TdF premiere. From then on, the three-litre Ferraris dominated from year to year; first the long-wheelbase models, then the short-wheelbase models, followed by double victories for the 250 GTO in 1963 and 1964. Most of the podium places went to Ferrari teams. The event then took a break until 1969, when Ferrari won again three years later, this time with the 365 GTB/4 Competizione. The triumphs of the Lancia Stratos equipped with a Ferrari Dino engine were followed by further Ferrari victories in 1981 and 1982; this time with the 308 GTB.

The history of the Tour de France, which lives on today as a classic event, ends in 1986. And as a model designation at Ferrari: the 250 GT with a long wheelbase was given the suffix TdF. It was followed by the Ferrari F12 TdF special model presented in 2015.

1972

1954

1959

At the time, Ferrari only produced cars to order for its illustrious clientele. Customers in a hurry sometimes received a half-heartedly tamed version of a racing car for everyday use. This also applies to the Thyssen family

WITH FLYING COLOURS

"Heini" Thyssen was one of the richest men in the world of his time. He was one of the great steel barons: The Thyssens rolled the steel through which industrialisation took shape; from factories, to railways, to cannons. As the shape of cities and the world of work evolves, the family business grows into an empire. Even in times of crisis, steel is an excellent way to earn money. And so, alongside the »old money« of the nobility, an equally wealthy »industrial aristocracy« emerges. This breathed life into the spheres of the new super-rich, introducing colourful and dazzling facets to their lives.

In the film »La Dolce Vita« (»The Sweet Life«), released in 1960, filmmaker Federico Fellini focuses on the decadent, supposedly carefree existence of »high society« in the 1950s. Rising from the ruins of the Second World War and benefiting from the rapidly advancing industrialisation of the years of the economic miracle with ever-increasing consumer choice, a new »high society« emerges, for which the term »jet set« is coined from the early 1960s onwards thanks to the emergence of private jet planes that transport them to almost every corner of the world in the shortest possible time.

As far as the Thyssen dynasty is concerned, the order of precedence is clear: The company founded by Heini's grandfather August is run by his son Fritz. His brother Heinrich (Heini's father) emigrates to Hungary. There he marries into the nobility. And »Heini« is eventually adopted by his father-in-law. This gives him the aristocratic title of baron. From then on, his full name was Hans-Heinrich Àgost Gábor Tasso Baron Thyssen-Bornemisza de Kászon. And his father's inheritance brings him immeasurable wealth.

In 1953, the Baron falls in love with Nina Dyer, a modle with British-Indian roots. She grows up in Ceylon (now Sri Lanka) on her father's tea plantation. But she is in search of the wider world. When she tries to realise her dream of acting in England, she is discovered as a model. Her next stop is Paris. At the side of fashion designer Pierre Balmain, the world of the rich and beautiful opens up for the attractive woman. This includes parties on the Riviera as well as intrusive photojournalists. The »paparazzi« follow the sweet life of the glamorous jet set at every turn. The associated tabloids document their goings-on between London, New York, Paris, Monte-Carlo, Capri and St Tropez. Some magazines describe Nina Dyer as »the best girlfriend a bikini ever had«. Hans-Heinrich Thyssen divorces her for her. And he showered Nina Dyer with gifts: these included the Jamaican island of Pellew Island, a coat made from the

The Gran Turismo sports cars usually wear the classic red colour of Italian racing cars. At Le Mans, they also have engines with a displacement of 4.1 and 4.5 litres. The one customised for Nina Dyer is a discreet three-litre. But with a top speed of up to 250 km/h, it remains a bullet on the road

Above the rear wheels of the 250 MM, the body designed by Pinin Farina has a horizontal ventilation duct. Hidden underneath is an opening to the wheel arch

Three carburettors supply the three-litre V12 with the right mixture of petrol and air. A dozen spare spark plugs can be seen on the side wall of the engine

fur of wild chinchillas, two black panthers (which sometimes wreak havoc in hotel suites), jewellery, art and the very Ferrari 250 MM that is now part of the Nationales Automuseum's collection.

At the time, Ferrari produced vehicles exclusively to order. Customers in a hurry sometimes receive a half-heartedly domesticated version of a racing car. This is also the case here. When Thyssen arrives in Maranello, the 250 MM is available ad hoc. But before the sports coupé (chassis number #0338MM; one of a total of 17, plus 14 spiders) is transformed into the dream car for Hans-Heinrich's future wife, it needs some work. It is painted at Pinin Farina in Turin in the grey-green Fiat colour »Grigio Dolomiti« – to match Nina Dyer's beloved Hermès trouser suit. Thyssen then has the interior fitted out directly by Hermès, including a tool bag made of the same grey deerskin as the collar of the aforementioned

The Berlinetta is one of Pinin Farina's early works for Ferrari. Initially, »Pinin« was still Battista Farina's nickname. From 1961, it became the new family name

The colour is originally found at Fiat. It perfectly matches the wardrobe of moel Nina Dyer

As is usual with early Ferraris, vehicle #0338MM also has the steering wheel on the right-hand side

trouser suit. In addition, the 240 hp coupé is fitted with glass panes instead of Plexiglas windows, additional rear lights and a heater.

Despite the lavish gifts, the marriage was short-lived. Shortly after the marriage, the young model had a fling with an even younger actor. This does not go unnoticed by Hans-Heinrich Thyssen. The divorce follows in 1956. His divorced wife is compensated with twelve million marks, a castle and jewellery. And the very next year, the model hooks the next big fish: Prince Sadruddin Aga Khan. In 1957, she married the »son of the religious millionaire Aga Khan« (as reported in the »Spiegel« magazine in 1956). His father's full name is Sultan Mahomed Shah Aga Khan. And in order to be allowed to marry a divorced woman, his son Sadruddin renounces the throne. His future wife converts to Islam and becomes Princess Nina Aga Khan upon marriage. The newly married couple enjoyed using the exclusive 250 km/h Ferrari 250 MM for excursions, for example to St. Moritz in winter. In addition to the coat of arms and decorations of the House of Aga Khan, it also has a roof rack mount – it is probably the only Ferrari racing car with a ski rack. She counters accusations that Nina Dyer has deliberately targeted wealthy men: »Luck fell on me without doing anything, it's been that way since I was a child. It just so happened that the big things in my life became my everyday life.«

»Heini« Thyssen has the interior made by the Parisian fashion specialist Hermès. It is made from the exact deerskin of the trouser suit that his young wife loves dearly

From the very beginning, Enzo Ferrari also knew his trade when it came to communication and marketing. A striking logo is just as much a part of this as the Cavallino Rampante

The five gear changes of the transmission can be shifted precisely in the shortest possible way using the filigree gear lever

Instead of sliding windows, the Thyssen-250, converted for road use, has hinged and cranked windows in the doors. Pockets swallow some of the luggage

The body of the 250 MM designed by Adriano Rabbone is straightforward. Every detail follows a purpose

For their wedding in August 1957, Prince Sadruddin presented his princess with a green Ferrari 250 GT Cabriolet (#0709GT). But even this marriage was short-lived: the couple separated in 1960 and divorced in 1962. Nina Dyer sells the Pinin Farina Cabriolet. In addition, a new settlement, now totalling 1.4 million US dollars, is credited to her account. The twice-divorced, very wealthy lady then orders a 250 GT Spider California (#4095GT) in Maranello. She donates the 250 MM to the ACO Museum in Le Mans. There, the similarly bodied 340 MM with a larger engine capacity played a major role in the endurance race. However, a sister vehicle of the Dyer-250-MM entered the list of winners of the 1953 Gran Premio dell'Autodromo di Monza, which was organised for sports cars.

In the end, Princess Nina Aga Khan's happiness did not last. Depression plagued her everyday life. In 1965, childless and aged just 35, she took her own life. She bequeathed her fortune to animal welfare. In the same year, Prince Sadruddin Aga Khan orders a silver-grey Ferrari 500 Superfast (#6049SA).

Enzo Ferrari is of the opinion: »A Ferrari is made for the racetrack, not for the boulevard«. The company founder's enthusiasm always centred on racing and the pursuit of even better, even faster racing cars. Secretly, the rich and beautiful, who increasingly ordered a Ferrari for everyday use, were met with incomprehension by Enzo Ferrari. However, the entrepreneur soon had to admit that his ambitious racing plans could not be adequately financed without the sale of exclusive one-offs. Ferrari needed at least three-digit production figures. And the three-litre twelve-cylinder came at the right time. From the mid-1950s; additional international sales opportunities opened up for high-performance sports cars. Over two dozen different models of the 250 series were produced, with an almost unmanageable number of different bodies. The luxurious Gran Turismo models were not affordable for everyone, but they found a large customer base that enabled Ferrari to continue to grow.

FERRARI 250 S

In 1952, seven Ferraris were already on the starting grid at Le Mans. Among the 212, 225 and 340 models, there was also a vehicle with the designation 250 for the first time, specifically a 250 S, which also appeared in other entries as the 250 Sport Berlinetta Vignale. The one-off, equipped with a new three-litre engine for testing purposes, made its racing debut at the 1952 Mille Miglia and did so in style: the Ferrari, driven by Ascari replacement driver and gentleman driver Giovanni Bracco, beat the favoured Mercedes team by four and a half minutes. Anecdote in passing: in addition to co-driver Rolfo, Bracco also has a high-proof companion in the car – it remains unclear whether he wins because of or despite the grappa that is within arm's reach...

At Le Mans, Ferrari and Mercedes-Benz clashed once again. This time, Alberto Ascari and Luigi Villoresi alternated at the wheel of the Ferrari test car. They took the lead right from the start and set the fastest lap of the race at 04:40.5 minutes and 173.159 km/h respectively. However, three hours after the start, they had to park the car with a burnt-out clutch. The next race for the 250 S with chassis number 0156ET, the 12-hour race in Pescara, brought another victory. Giovanni Bracco and Paolo Marzotto were once again at the wheel of the »Experimental GT«. At the end of the season, the Carrera Panamericana is another motorsport highlight on the programme. Bracco once again duels with the competition from Stuttgart. He dominates the race, winning the first seven of a total of eight special stages in this adventurously fast race. But before the start of the eighth stage, Bracco let the Mercedes team know that they could minimise the risk and slow down; he would retire in the final stage. Mercedes inherited an important victory for the 300 SL and the brand.

FERRARI 250 CALIFORNIA

In the film »Ferris Bueller's Day Off«, a California Spider (with the US licence plate »NRVOUS«) plays a central role – although a total of three replicas are used in the film. The high school comedy from 1986 is set in Chicago. And America is an important setting for the story of the open-top Ferrari. After all, it is the US importers who tell European car manufacturers about the great popularity of open-top sports cars and thus pave the way for corresponding models. In the case of Ferrari, these are US importer Luigi Chinetti and West Coast-based John von Neumann. As a result, Ferrari soon added the 250 Spider California, manufactured by Scaglietti, to the luxuriously equipped 250 GT Cabriolet Pininfarina, which was produced in a small series.

Three series of this model were produced, totalling 106 units, both on the long wheelbase of 2600 mm (1957-1959) and on the »passo corto« of the 250 GT SWB with 2400 mm (1959-1962). The engine output varied – depending on the development stage of the V12 – between 240 and 280 hp. The top speeds quoted at the time were a breathtaking 252 and 268 km/h respectively. Nine »California« – as this model is also known for short – were even produced with a lightweight aluminium body. They were intended for sports use, for example at Le Mans, Sebring or the 1000-km race on the Nürburgring.

FERRARI 250 GTE

Enzo Ferrari set his coachbuilder the task of creating a four-seater coupé for the 166 presented in Turin in 1948. In view of the large engine and the 85-litre fuel tank, this was quite a tricky task, as Touring boss Bianchi Anderloni later recalled. Twelve years later, Ferrari launched the 250 GT 2+2, its first mass-produced four-seater.

With a length of 4.7 metres, the coupé has quite an imposing appearance. Ferrari's in-house couturier Pinin Farina ensures that it is also an elegant car. The car impresses with its brand-typical driving dynamics while offering space and luxury. 240 hp and a top speed of 220 km/h go hand in hand with space for four people and luggage. This is also made possible by the twelve-cylinder engine, which has been moved 17 centimetres forwards in the tubular frame chassis. Behind the wooden Nardi steering wheel is the lavishly equipped dashboard, with seven horizontally arranged round instruments. At the time, »auto, motor und sport« attested: »With three people on board and a full tank (almost 1800 kilograms), this Ferrari can accelerate from 0 to 160 km/h in 25 seconds and brake to a standstill again. This measurement says everything about the all-round capabilities of such a fast, heavy vehicle.« Incidentally, the brakes are now disc brakes. »Only if you want to race should you buy the Berlinetta from the same company. Otherwise, there's nothing better than this 2+2, which is as desirable and fast as it is expensive.«

To that end, Ferrari offered 50 visually identical examples of the 250 GT 2+2, sporting a four-litre V12, at a cost of DM 53,000 (Deutschmarks) each. They were given the designation 330 America and marked an important milestone in the development history of Ferrari's large Gran Turismo models. Incidentally, a Ford Taunus 17 M – the »bathtub« – cost DM 6950 at the time.

»What I wanted for my cars,« says Enzo Ferrari, »was character.« In Battista »Pinin« Farina, he found a kindred spirit and the right designer. It was not until 2010 that Ferrari founded its own design department

NEW CLOTHES FOR THE KING

Giovanni Battista Farina was born in 1893, the tenth of eleven children in a Piedmontese family. There he was nicknamed »Pinin«, affectionately meaning »the little one« or »the baby«. When Battista was twelve years old, his brother Giovanni, who was nine years older, set up his own business. In Turin (at Corso Tortona 12), he founded the Società Anonima Stabilimenti Industriali Giovanni Farina, or Stabilimenti Farina for short. Battista and his other brother Carlo are involved right from the start. »Pinin« began as an apprentice. At the age of 17, he was already head of design at the family business, which he left twenty years later to set up his own business, Carrozzeria Pinin Farina.

Thanks to the takeover of Carrozzeria Casaro, which had run into difficulties in the wake of the global economic crisis, »Pinin« also had suitable workshop premises at Corso Trapani 107 in Turin. Incidentally, the predecessor company Casaro made a name for itself with its first self-supporting bodywork for the Lancia Lambda.

The family business Stabilimenti Farina is also experiencing difficult times: No suitable successor for the company founders has been found. Giovanni's son Nino pushed ahead with his successful racing career and his brother Attilio was unable to lead his father's company into the new automotive age. Stabilimenti Farina only existed until 1953, but »Pinin« had long since become a big name...

The year before saw the start of Pinin Farina's collaboration with Ferrari. The first vehicle was an ivory-coloured 212 Inter Coupé with a black roof. Initially, however, Pinin Farina was one of many suppliers. Allemano, Bertone, Boano, Farina, Fontana e Reggiani, Ghia-Aigle, Ghia, Touring and Vignale also clad the Ferrari, which were produced in small numbers. But Ferrari liked Pinin's clear designs. »After trying out all the coachbuilders,« writes Pete Lyons, »he decided that Battista ›Pinin‹ Farina was the best couturier for his cars«. Enzo Ferrari later called Battista Farina the »unique one«. »Pinin brought about the renaissance in Italian car design,« says Ferrari. »His car bodies are simple and clearly drawn, reduced to the essentials, so to speak.«

The two men speak the same language. But it took the skilful mediation of Battista's son Sergio to get to this point. Battista is just as stubborn as Enzo: it is out of the question for Ferrari to visit Farina in Turin, just as Battista Farina refuses to go to Maranello. »La Stampa« journalist Gino Rancati feels reminded of »two incomparable kings«. Nevertheless, the two knew each other: Three decades earlier, they had met repeatedly as participants in the Aosta-Gran San Bernardo race. Farina on Itala (1921) and Fiat 1500 Corsa (1922) in the Turismo category, Ferrari on Alfa Romeo and Steyr in the sports and racing car class.

After the Dutch Prince Bernhard receives a bespoke Pinin-Farina design, the Belgian Princess Lillian von Réthy has this Fuoriserie model uniquely tailored for her

After Sergio Farina arranges a meeting halfway through 1951, things move quickly. Further orders followed in quick succession. »Our interests complemented each other perfectly: he was looking for a beautiful woman to dress, and I was looking for a good tailor for my beautiful wife,« Enzo Ferrari recapitulates in his usual vivid manner. Later, the two entrepreneurs were united by a deep friendship based on respect. It was one of the few that Enzo Ferrari cultivated. Battista entrusted his son Sergio with

Princess Réthy and her husband, the Belgian King Leopold III, used to reach into this steering wheel

The coupé's body appears opulent thanks to the sweeping wings that extend far beyond the track at the sides

Spoked wheels with a central securing spinner have always been the norm at Ferrari

The distinctively designed watches are from Veglia

responsibility for Pinin Farina's Ferrari designs and the contact between the two companies' technical offices – today this is known as a key account.

Until 6 June 1961, the company is still called Pinin Farina before the Italian state accepts Pininfarina as the new family name by decree – an unusual event even in Italy. It is a state thank-you for the fact that Pininfarina's creations have long since become a global figurehead for Italian design and Italian industry.

As far as design development is concerned, Pinin Farina experiments with the help of a series of individual pieces, so-called »Speciale«. This makes it easy to try out new lines, design elements and details. »To stand out from the crowd, you need at least one ›Fuoriserie‹: special bodies in limited editions,« explains design professor Paolo Tumminelli. »But the jet set prefers the ›Esemplare Unico‹, the one-off.« Just like the lady in Paris with haute couture, the gentleman has his car dressed in Turin. According to Tumminelli, the influence of the »fuoriserie« phenomenon on the development of automotive design is often overlooked. »The customers' taste and style are beyond any doubt; after all, these people are as at home anywhere in the world as they are in art, architecture and fashion. They are eccentric and keen to experiment. They commission the best coachbuilders in the world, but play designer themselves...« In this way, design ideas can be tested before they possibly find their way into series production – »...like a kind of market research«.

The Ferrari 375 Plus Pinin Farina Cabriolet produced for the Belgian king in 1955, for example, was the first to feature elongated Plexiglas covers over the headlights. This element would become typical of numerous sports cars in the following years – from the 250 GTO and Jaguar E-Type to the Alfa Spider. With the jet age, the cars also grew tail fins and other wing-like elements. They became standard in the USA. Pinin Farina's Alfa Romeo 6C 3000 CM Super Flow Coupé Sperimentale from 1956 and the Ferrari 410 SA Superfast Coupé Speciale from the same year are among the pioneers of this fashion.

The Coupé Speciale (#0751GT) from the Loh Collection was created in 1957 at the request of Lillian Princess von Réthy, shortly after the Dutch Prince Bernhard had been delivered a similar vehicle (#0725GT) with a three-litre V12 and the later look of the 410 Superamerica. The princess was formerly a nanny at the Belgian court before marrying the Belgian King Leopold III in 1941. Six years earlier, he had lost his wife Astrid in a road accident in Switzerland. King Leopold III was not without controversy in Belgium, partly because of his choice of second wife,

Ferrari's fashion designer Pininfarina knows how to stylishly incorporate and emphasise details. Like these skilfully crafted openings in the vehicle flanks

The »short« twelve-cylinder engine was designed by Gioacchino Colombo. The spark plugs are located between the two cylinder banks. The three carburettors and the air filter box are enthroned above them

Spotlights in the depth of the radiator opening become characteristic details of the great Ferraris of the time

While the front appears impressively in the rear-view mirror, the rear remains a rather plain image in the memory of overtaken drivers

Another stylistic feature, which first appeared in the latter half of the 1950s, are the headlights behind Plexiglas covers

The discreet tail lights at the tip of the fin-like wings are one of the unique features of this one-off – Fuoriserie in fact

but even more so because of his dealings with the Nazis during the Third Reich. The king spent years in exile and was ultimately forced to abdicate in 1951. Instead, his son Baudouin assumed the throne at the age of 21. Meanwhile, Leopold had a bone to pick with Enzo Ferrari – and not just as a good customer: when Dunlop ended its long-standing partnership with the Scuderia in 1955, the ex-monarch found a valuable replacement in the Belgian tyre manufacturer Englebert.

Pinin Farina himself celebrated the unique grey-painted car with its natural-coloured leather interior, which was delivered in January 1958, as a combination of uniqueness and simplicity. He was happy to use it for advertising purposes. The car shows features of a 250 GT design developed nine months earlier. What is new is the emphasis on the outline of the rear wings with the tail lights integrated on top. The all-round rear chrome bumper is clumsy by Ferrari standards. At the front, four buffer-like chrome elements protect the aerodynamically shaped

Enzo Ferrari: »What I wanted for my cars was character. And I found it with the help of Giovanni Battista ›Pinin‹ Farina«

front with the headlights positioned under Plexiglas covers. Two further headlights are concealed behind the traditionally shaped radiator grille. The instrument arrangement is as lavish as usual: the speedometer and rev counter are full-sized. They are flanked in the centre of the dashboard by five smaller instruments for water and oil temperature, oil pressure, fuel level and time.

Battista Farina knows how to move with the times: He handed his company over to his son Sergio and son-in-law Renzo Carli in good time, who in turn set the course for the transition from manufacture to industrial production without losing sight of the avant-garde. At the time of the delivery of the 250 GT Speciale to the Princess, Pinin Farina was preparing to move to Grugliasco. Just outside Turin, there is space for modern production facilities. An R&D centre was added in 1966 and the first wind tunnel built for the automotive industry in 1972. Pininfarina continues to produce series cars and one-offs, which continue to provide valuable impetus for automotive design.

CARROZZERIA TOURING

Superleggera – Carrozzeria Touring made a name for itself worldwide with this patented construction method. The name is both an accolade and a mark of distinction for the vehicles, which the Milan-based coachbuilding specialist provides with a lightweight aluminium skin stretched over a filigree tubular lattice construction. As in aviation, the coachbuilders focus on aerodynamics and lightweight construction. Among the most famous representatives with a Superleggera body are numerous Alfa Romeo, Lancia, some BMWs, Aston Martin, Maserati, the first Lamborghini and both the Auto Avio Costruzioni 815 and early Ferraris. Between 1948 and 1956, Touring built a total of 87 166, 195, 212, 340, 500 TR and 625 LM roadsters for Maranello.

The history of the company began in 1926 with the takeover of Carrozzeria Falco. The Touring founders Felice Bianchi Anderloni and Gaetano Ponzoni were lawyer friends. Until the two lawyers swapped the law books for business books, the Falco company belonged to Vittorio Ascari, the brother of the successful racing driver Antonio Ascari – one of Enzo Ferrari's few idols.

The Carrozzeria Touring caused a sensation with designs such as the »Flying Star« from 1931, an elegant, snow-white body on the chassis of the Alfa Romeo 6C 1750 GS and Isotta Fraschini 8B. Location of the event (as was customary at the time): Concorsi d'Eleganza, as organised in the garden of the Villa d'Este on Lake Como.

In 1966, the company goes bankrupt due to a major order cancelled by the British Rootes Group.

CARROZZERIA ZAGATO

Ugo Zagato gains his first experience in lightweight construction and aerodynamics with the production of aeroplanes used by the Italian Air Force in the First World War. The fighter pilot Baracca, whose coat of arms Enzo Ferrari would later use as his trademark, flew one of these Ansaldo aircraft.

In 1919, Ugo Zagato founded his coachbuilding company for automobiles near Milan and quickly became a couturier for Alfa Romeo. His lightweight bodies were soon used in motorsport. The six- and eight-cylinder Alfa cars of Scuderia Ferrari are almost exclusively bodied by Zagato. Despite all its sporting successes, Zagato repeatedly proved to be economically unstable. As a result, the company experienced various start-ups and also a period in which Zagato's creative ideas were realised in sporty coachwork under the name Brianza. These included the versions of the 8C that were successful at the Mille Miglia, among others, but also the Grand Prix model that became world famous under the name Monza. The aluminium skin of the »Alfetta« was also created in Zagato's workshop.

Towards the end of the 1930s, the streamline increasingly determined the shape of the extremely lightweight bodies. These continued to be produced either at the customer's request or – on behalf of the vehicle manufacturer – as one-offs or in small series. Examples include the Ferrari 166 MM or the six 250 GT Zagato, which were created in the mid-1950s and – like many Zagato designs – were intended for racing. The two bulges in the roof are a characteristic feature of the 250 GTZ; the double bubble creates space for the driver's and passenger's helmets.

The one-off 575 GTZ (2006) and 599 GTB Nibbio Zagato (2008) continue this tradition.

CARROZZERIA SCAGLIETTI

Enzo Ferrari and Sergio Scaglietti, 22 years his junior, are well acquainted. Since the 1960s, the two Modenese have been meeting regularly on Saturdays to eat and chat together. The two share a passion for fast, red cars and have known each other since 1937, when, as a teenager, the young Sergio began fixing up Scuderia Ferrari's battered Alfa Romeos returning from racing battles. Scaglietti developed great skill and know how enabling him to shape entire car bodies according to his eyes and instinctive touch.

In 1951, the coachbuilder founded his own Carrozzeria, now in the neighbourhood of Ferrari in Maranello. He repaired damaged Ferraris, maintained contact with the company and got on well with Enzo's son Dino. It was from him that the impetus came to entrust Scaglietti with the construction of racing cars.

What began with a 500 TR developed into a permanent fixture for the rapidly growing company. The 250 GT Berlinetta, 250 GT Spider California and 250 GTO models marked the high points of Scaglietti's work. Series production of numerous models designed by Pininfarina was also soon realised. In the 1970s, Scaglietti was taken over by Ferrari and from then on officially functioned as the sports car manufacturer's body plant. Scaglietti is well utilised with the production of the eight-cylinder models – the 308s coming off the production line at Scaglietti soon represents three quarters of Ferrari's entire annual production.

In 2004, Ferrari paid homage to its coachbuilder with the 612 Scaglietti. Today, the »Carrozzeria Scaglietti programme« stands for comprehensive customisation options, far beyond paint and leather.

»For those hungry for experience, it enables driving in itself and prepares one for the impossible,« enthuses Heinz-Ulrich Wieselmann in »Motor Revue« about the Ferrari 250 GT SWB, which Wolfgang Seidel made available to »auto, motor und sport«

ANYTHING BUT A JOKE

... in 1960, and »Motor Revue«, which is published quarterly by the same publisher. The Ferrari »is loud, but easy to drive and as willing as a summer sweetheart.« A phrase that today – seven decades later – is of course just as inappropriate as film titles such as »00Sex am Wörthersee«.

In the post-war reconstruction years of Germany, however, people were not bothered by such blunders. Instead, the film title stands for a three-digit number of productions that were made at the time with the Austrian actor Gunther Philipp. They first appeared on the cinema screen and later also on television. Gunther Philipp was one of the busiest German-speaking film and television actors of the 1950s and 1960s. He often played the role of the »fool«. He acted alongside Peter Alexander, Waltraut Haas, Hans Moser, Vivi Bach, Hans-Jürgen Bäumler and Peter Weck. The films, which were released in rapid succession between 1949 and the 1970s, have titles such as »Kleiner Schwindel am Wolfgangsee«, »Kaiserwalzer«, »Die Abenteuer des Grafen Bobby«, »Der fröhliche Wanderer« and »Lügen haben schöne Beine«. He is one of the great stars of his time. However, his professional background and sporting side are hardly known to the public.
Gunther Philipp is the pseudonym of Dr Gunther Placheta, a qualified psychologist and doctor. As a teenager, he was a member of the Austrian Olympic squad; he held an unbeaten record in the 100 metre breaststroke for 14 years. In 1959, he founded his own motorsport

With its design by Pininfarina and realised by Scaglietti, the Berlinetta became the archetype of the classic Gran Turismo. As a Competizione, with a lightweight aluminium body, this model is a rarity. Its black colour scheme makes it a unique and exotic example

The 250 GT SWB is regarded as the ideal type of Gran Turismo. It is ideal for covering long distances at speed, as well as for successful racing

An extensive collection of clocks in the centre of the dashboard complements the two round instruments behind the steering wheel

team with the »Ecurie Vienne« and made his debut with two Formula Junior racers and a Mercedes-Benz Gullwing. »In eight years of active motorsport,« says Philipp, he contests »eight to ten races a year as an amateur«. »Constantine«, »Carlo Brioni«, »Terramotte« and »Giulio Pavesi« are other pseudonyms used by the gentleman driver, who competes exclusively in Austria.

The Mercedes 300 SL is followed by a Ferrari 250 GT. But before his own SWB (#2839GT) is delivered to the actor, the German businessman and racing driver Wolfgang Seidel lends him his 250 GT Competizione (#1807GT).

Philipp wins the »Prize of Vienna« in 1961 and becomes Austrian national champion in the GT class. In the years

A trough sits above the carburettors, from which the stacks extend. Coarse-meshed grilles are designed to prevent foreign bodies from getting into the combustion chambers

The Italian tricolour is part of the Ferrari logo, the German colours adorn the Seidel SWB

Race track or red carpet — the 250 GT can do both. Today, the cage and emergency stop are mandatory

Five gears are also part of the standard repertoire in the 250 GT SWB

that followed, the racing physician/actor managed to defend his title twice, now at the wheel of his own Ferrari.

Gunther Philipp's metallic grey 250 GT bears a stripe in the Austrian national colours running down the middle of the vehicle. On Seidel's black-painted GT, the stripe shines in black, red and yellow. Wolfgang Seidel was born in Dresden in 1926 and lived in Düsseldorf after the war. There he ran the company Autosport Wolfgang Seidel GmbH, which dealt in exclusive and exotic cars. From 1950, Seidel became active in motorsport. Between 1953 and 1962, Wolfgang Seidel competed in the premier class a total of ten times. His cars: Veritas, Maserati, Cooper-Climax and Lotus. Best result: ninth place in the 1960 Italian GP. In 1955, he and Olivier Gendebien finished fifth at Le Mans in a Porsche 550 Spyder.

In 1958, he shared the works Ferrari 250 TR 58 with Wolfgang Graf Berghe von Trips. Wolfgang von Trips em-

bodied – as »auto, motor und sport« wrote at the time – »a mixture of confidence, recklessness and ambition«. The »ams« wrote about his team-mate: »There is no doubt about Wolfgang Seidel: he is certainly not as fast as Trips, and nobody believes that he will ever be able to exceed his and the vehicle's limits. Seidel is considered a veteran.« But things were to turn out differently in 1958: The two were in first to third place for hours in the race, which started in bright sunshine and then experienced heavy downpours. Shortly after midnight, Seidel skidded off the track at Arnage and was stranded in a pile of sand, there to protect the spectators from cars sliding out. Ferrari team-mates Hill and Gendebien won the race. In the two years that followed, he entered the race with Edgar Barth in a Porsche 718. Seidel and his team-mate were the clear front runners in both races. But in 1959, gearbox damage ended their race after 168 laps; in 1960, they took class victory and eleventh place overall. The Targa Florio victory in 1959 and two third places (Caracas 1957 and Sebring 1958) stand out among the World Sports Car Championship races that Seidel contested. In Sicily and Florida, Seidel was once again on the road with Barth in the works Porsche, while in Caracas the Düsseldorf native once again shared the Testa Rossa, named so by the factory, with Wolfgang von Trips. From 1960 onwards, the black 250 GT Competizione was Seidel's favourite sports car. He contested numerous circuit and hill climb races with it for three seasons, always finishing on the podium, usually at the top.

Before the end of the 1960s season, Seidel had the original engine fitted with carburettors from a 250 TR in Maranello. While the twelve-cylinder engines in the competition coupés were rated at 280 horsepower, the Testa Rossa had an impressive 300. The 1961 season also started with such an engine, with Seidel handing over his »Passo Corto« to Gunther Placheta, alias Philipp. In the following year, Wolfgang Seidel then entered the car three

The carburettors come from the specialist Weber. Different outputs can be generated with different sizes

While the majority of Scaglietti's 250 GT models, which are already powerful in production trim, are clad in sheet steel, the Competizione is clad in lightweight aluminium

In the early 1960s, aluminium wheels increasingly appeared on the racing horizon. Ferrari sticks with wire wheels

times for Formula 1: in Holland, Great Britain and for the German GP. The latter became a turning point in the gentleman driver's racing career: due to a technical defect in his Lotus-BRM, Seidel only completed four of the five prescribed practice laps. And although Seidel can already point to numerous Nürburgring successes – and therefore in-depth knowledge of the circuit – he is not allowed to start. This displeases Seidel, who then publicly criticises the organiser. The organiser then withdrew his AvD membership, which resulted in the loss of his racing licence.

»Young, impetuous and full of excess power« is how »auto, motor und sport« describes the three-litre coupé that Seidel makes available to the publisher for test drives. The 250 GT Berlinetta Passo Corto celebrated its world premiere at the 1959 Paris Motor Show. In keeping with its name, the 250 GT SWB had a wheelbase that was 20 centimetres shorter than the previous 250 models. Under its body, designed by Pininfarina and manufactured by Scaglietti, disc brakes and hydraulic telescopic shock absorbers were used for the first time in a production Ferrari. The short wheelbase ensures a modern appearance. Above all, it had a positive effect on the handling of the Gran Turismo, which dominated the international GT scene for three years – until the appearance of the 250 GTO, which was optimised above all in terms of aerodynamics. The 250 GT SWB is archetypical of the »Grand Turisme« theme: a sports car with space for passengers and luggage, equally suitable for the racetrack and the road. Between 1959 and 1962, a total of 165 examples of this Berlinetta were built, hardly any of which were exactly the same as any other, as they were all handcrafted. The model also underwent a series of modifications. And then there are the differences between the road-going version and the 20 Competizione racing cars. The latter have an aluminium body. Carpets, insulation material and bumpers are omitted, as are other equipment details that Ferrari customers will appreciate in everyday use. Overall, the competition GTs weigh 110 kilograms less than their steel series siblings. And there are also differences under the bodywork: While the fuel tank of the racing coupé holds 130 litres, the standard trim has 90, which creates space in the boot. And with 220 to 240 horsepower, the road-going Ferraris are already generously motivated; whilst the racing units produce between 260 and 280 hp. A special feature of the twelve-cylinder engine is the new cylinder heads with externally located spark plugs. This arrangement, first used on the Testa Rossa (250 TR), enables a new combustion chamber geometry and creates space in the engine V for three quadruple carburettors.

Wolfgang Seidel once advertised his Berlinetta in »auto, motor und sport". His erstwhile GT, which impresses with its high degree of originality, has been part of the Loh Collection for more than two decades.

ASTON MARTIN DB4 GT ZAGATO

The GT Zagato is actually better suited as a counterpart to the Ferrari 250 GTO, as it is – like the »Gran Turismo Omologato« – the evolutionary model of a series predecessor. However, the proportions and appearance of the British car, which Zagato has moulded into shape, make it an obvious choice.

Between 1961 and 1962, Carrozzeria Zagato built a total of 19 vehicles of this model based on the Aston Martin DB4 GT produced in Newport Pagnell in the UK. In typical Zagato fashion, the Italian bodywork specialists focussed on weight and aerodynamics. As a result, the DB4 GT Zagato weighs two and a half hundredweights less than the DB4 GT, which was already designed for GT racing. A good part of the weight advantage is due to the Perspex windows, which retain their shape even at high speeds. The 3.7-litre in-line six-cylinder engine, with dual ignition, mobilises 314 bhp; though the actual figure is around 285 hp. This enables the Gran Turismo to accelerate to 60 mph (around 96 km/h) in 6.1 seconds and, with the usual rear axle ratio, to a top speed of 247 km/h.

Ercole Spada, the 23-year-old head of Zagato design at the time, was responsible for the looks of the evolutionary model. After his time at Zagato ('60-'69), he would continue to attract attention well into the 1990s with successful body designs, including the BMW 5 Series and 7 Series (E34 & E32) as well as the Alfa 155.

Aston Martin charged £5,400 for the Zagato at the time. The DB4 GT intended for GT sport costs around £4600 and the standard DB4 is listed at just under £4000. By comparison, a Mini Cooper cost £500 at the time. As far as the DB4 GT Zagato's racing career was concerned, it was overshadowed by Ferrari. Even more so when the 250 GTO appeared in 1962, becoming a rival that would prove difficult to beat.

JAGUAR E-TYPE

When Jaguar founder Sir William Lyons presented the Jaguar E-Type at the Geneva Motor Show in March 1961, his company could look back on five Le Mans successes. Scuderia Ferrari only had three victories in its books. The E-Type becomes the star of the show. The media talk about the »most fantastic toy for adults ever presented at a motor show«.

For Enzo Ferrari, the elegant coupé is quite simply »the most beautiful car in the world«. And Ferrari sales boss Girolamo Gardini feared that the E-Type could follow in the footsteps of the D-Type and that Ferrari would have to admit defeat to the new GT from Coventry on the racetrack. According to Giotto Bizzarrini's recollections, the Jaguar E-Type gave Enzo Ferrari the impetus to develop the 250 GTO. He then commissioned Bizzarrini to realise this project.

For Jaguar, on the other hand, the E-Type's modest sporting endeavours remained the preserve of private teams. The situation was different on the road: The E-Type was a complete success. By the end of production in 1974, 72,500 units had been built, initially with a six-cylinder engine and later with a twelve-cylinder engine. Customers received a powerful sports car with a nominal output of at least 265 hp and a top speed of well over 200 km/h. And all this at an extremely attractive price.

The media write of a »dream car for everyone – at democratic prices«. In »auto, motor und sport«, racing driver and car tester Paul Frère specified in 1962: »Until then, all road-going cars with a top speed of around 220 km/h or more – the ones we used to call the »really fast ones« – cost at least around 40,000 marks. Can it be possible that a car comparable to these supercars is offered for about half that price in its country of origin?«

BIZZARRINI GT STRADA 5300

The just 1.1 metre tall sports car, with a front-mid-engine design, appeared on the scene in 1963 as the Iso A3/C (C for Corsa). It was the dissimilar sister model to the Iso A3/L (Lusso) known as the Iso Grifo and the result of a collaboration between Giotto Bizzarrini and the entrepreneur Renzo Rivolta.

Iso became famous in Italy for the production of motor and cabin scooters, and BMW also produced the Isetta under licence in the 1950s. And after the enthusiastic motorboat racer Rivolta was repeatedly annoyed by his Jaguar and subsequently his Maserati 3500 GT, he decided to produce his own vehicle in the upper market segment. The result is the Iso Rivolta, a two-door coupé with a body designed by the newly appointed Bertone design chief Giorgetto Giugiaro and an American V8 engine. Iso thus combines performance with reliability and manageable costs.

The A3/L and A3/C also combine Italian chic with the small-block V8 originally intended for the Chevrolet Corvette with a displacement of 5354 cm^3. While the A3/L is also designed by Giugiaro, Bizzarrini is not only responsible for the design, but also – in co-operation with Bertone, as it is said – for the lines of the Iso A3/C. »I started with the idea of the 250 GTO and tried to improve it,« recalls Giotto Bizzarrini, who worked first at Alfa Romeo and then at Ferrari after completing his studies. When he left Maranello in 1961 as part of the »palace revolution«, he designed the V12 engine for Lamborghini's first sports car (3500 GT), was involved in the underfunded ATS project and realised the sports car ideas for Iso. From 1965, after differences of opinion with Renzo Rivolta, the sports car, which could reach speeds of up to 275 km/h, was marketed under his own name as the Bizzarrini GT 5300.

A Ferrari is always something special, and limited edition models are the pinnacle of this. The crowning glory are one-offs, such as the Speciale creations from Pininfarina at Ferrari. Among connoisseurs, this is called Fuoriserie

EXCLUSIVE ELEGANCE

»Aerodynamics iare for people who can't build engines.« This quote is also attributed to Enzo Ferrari. And indeed, Ferrari engines are traditionally considered superior to their competitors. Ferrari's success is based above all on the strength of its engines. Added to this is a traditionally stable tubular frame chassis that offers good handling characteristics. However, in the 1950s, streamlined sports cars à la Panhard increasingly demonstrated the importance of efficient aerodynamics. Despite their extremely modest engine power, they achieved top speeds of almost 200 km/h at Le Mans. Despite its 280 hp, the Ferrari 250 GT SWB only reached 250; like some Ferrari models of its time, it was far from being able to convert its power into speed in the best possible way...

Knowledge of aerodynamics is not new. Pinin Farina, Touring and Zagato were already creating streamlined, optimised cars in the pre-war period. Initially, however, this had little significance for series automobile construction. But as performance levels rose and the number of vehicle models increased, the topic of aerodynamics once again became a focal point. All of a sudden, aerodynamically designed vehicles became fashionable. They epitomise performance and speed. At the same time, superlative model names embodying the outstanding position of these vehicles became fashionable: at Ferrari, the »Super GT« genre emerged, with models such as the Superfast and Superamerica.

A striking detail of the rear – in addition to the back line – is the integration of the rear lights into the bumper elements

They are children of a seemingly carefree time. The war was long over and the economic miracle was followed by a never-ending boom. »Rise without end« is the title of the CDU advertising film that accompanies Chancellor Konrad Adenauer into another term of office. The five millionth Beetle rolls off the production line in Wolfsburg. And records are also celebrated in Turin. Everyone is doing well. »These are golden years for motorists: money and petrol are plentiful, there is no speed limit on the roads. Baby boomer families want to have fun fast,« summarises design professor Tumminelli.

The designation Superamerica (SA) characterises the Ferrari models with the largest displacement. They are primarily intended for customers on the other side of the pond. The 410 model (with a 4982 cc Lampredi twelve-cylinder engine with between 340 and 360 hp) from 1955 was followed four years later by the 400 SA, also with 340 hp. Its four-litre V12 was based on the »smaller« version developed by Colombo, which had celebrated its premiere in the 125 S twelve years earlier. In view of their performance and their sensational looks, the Speciale, aerodynamically clad by Pinin Farina, were given the appendix Superfast. In fact, they enable their owners to reach top speeds well in excess of 250 km/h. The rear design of the Superfast is striking: it slopes in a flat, straight line from the roof over the full-size rear window to the bumper halves with skilfully integrated tail lights. On the Superfast Coupé Spécial Aerodinamica, the rear wheel arches are sometimes also covered with so-called »spats« – as on the Jaguar XK 120.

The fact that the topic of aerodynamics also came to the fore in Maranello is thanks to the engineers around Carlo Chiti and Giotto Bizzarrini. And Paul Frère. The journalist and racing driver confronted Enzo Ferrari with the low top speed of his LM winning car from 1960: despite over 300 hp, the Belgian stopped his 250 TR 59/60 at a maximum of 259 km/h. Chiti, Bizzarrini and the very young Mauro Forghieri were responsible for the technical development at the time. However, a large part of their task consisted of convincing Enzo Ferrari of the advantages of new approaches. For example, the introduction of disc brakes and the mid-engine design were among the fruits of their labour, which put the Scuderia back on the road to victory. The highlights of their work include the mid-engined Formula 1 car F156 »Squalo«, the 246 SP, also equipped with a mid-engine, and the 250 GTO.

A 250 GT Berlinetta Pininfarina Sperimentale Le Mans with a 300 hp TR engine, aerodynamic front and elongated, sloping back accompanies the GTO. Ferrari entered the car at Le Mans

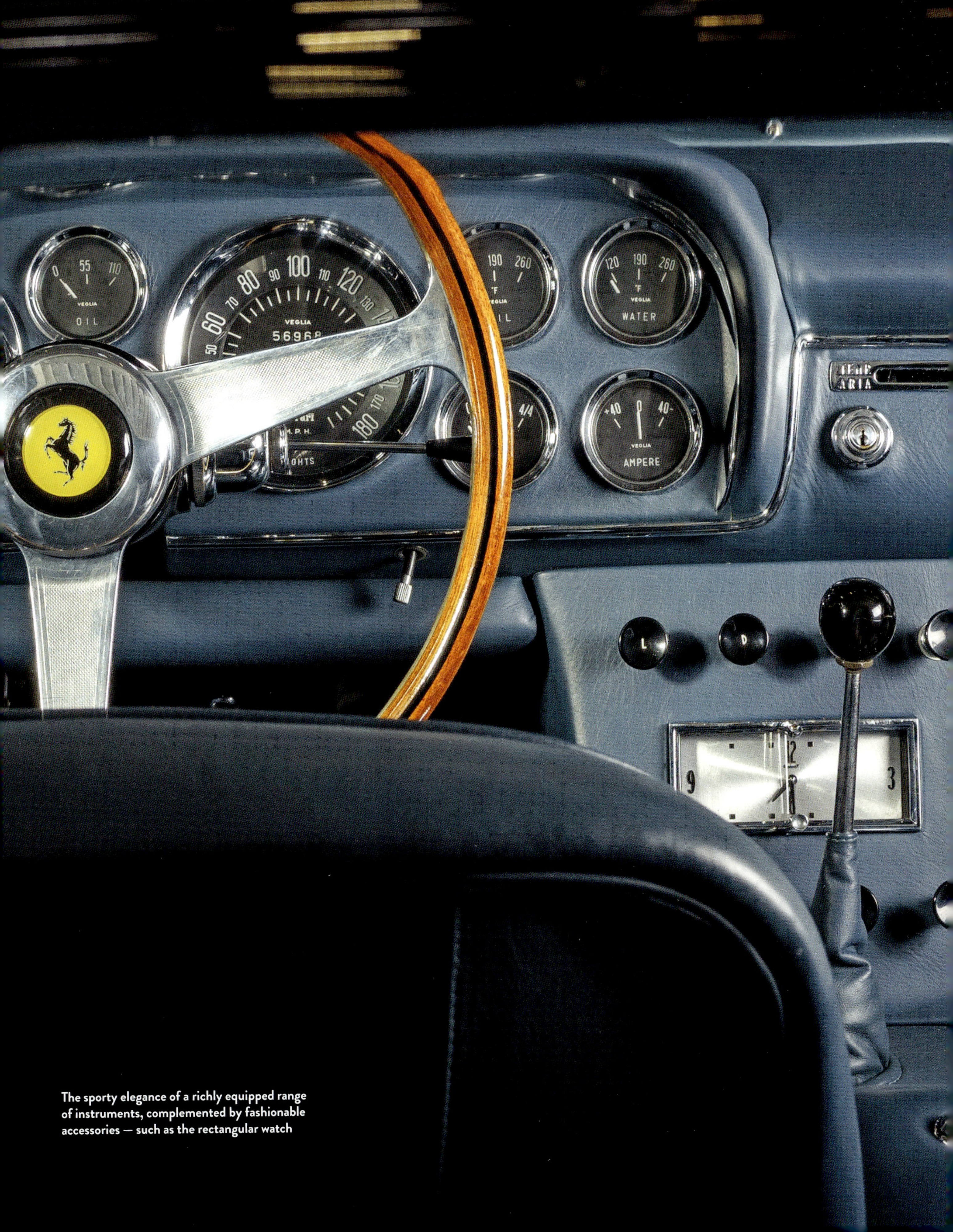

The sporty elegance of a richly equipped range of instruments, complemented by fashionable accessories — such as the rectangular watch

The flat, sloping silhouette of this one-off is also found in the Superfast. It epitomises aerodynamics – without actually being aerodynamically optimal

in 1961. At this time, the company had already been trading as SEFAC (Società Esercizio Fabbriche Automobili e Corse) for a year. Battista Pininfarina is one of the members of the Board of Directors. Two days after the 24-hour race in France, Ferrari delivers chassis #2821GT to Pininfarina. The body of the sky-blue (Celeste metallizzato) 250 GT Speciale of the Nationales Automuseum is built in Grugliasco on the basis of this chassis. Pininfarina originally intended to build a Cabriolet Speciale – but the coachbuilders quickly changed their minds. And so this unique Coupé Speciale was created. It was one of four »Berlinetta Esperimentale« realised by Pininfarina. The front of the car is reminiscent of the 250 GT Pininfarina,

The metallic particles give the light sky blue a unique lustre and the vehicle a sense of lightness

A rev counter that reaches up to 8000 revs is just as exclusive as a speedometer with a scale that counts up to 180 miles per hour. That corresponds to 290 km/h

The simple design of the radiator grille is characteristic of the sports cars from Maranello – right from the start

its rear of the 410 SA. The sky-blue coupé is a highlight of the Motor Show held at Earls Court in London in October 1961. The car gave visitors to the show a foretaste of the new Ferrari 250 GTO to be launched in 1962.

The hand-picked Speciale proved to be a stroke of luck for Ferrari and Pininfarina. On the one hand, the one-offs allow design elements to be trialled. If they prove popular, some of them are later incorporated into the series. And on the other hand, Ferrari satisfies the needs of its most solvent customers – for individuality and exclusivity beyond the series products.

The vehicle on display at the London show will be sold in November via the British importer Maranello Concessonaires to a compatriot who will use it at his residences in Jersey and Portugal. It then ends up in the hands of various collectors, including the Texan Mecom family. They have been regular customers in Maranello since 1957. Their money came from the oil business. Incidentally, they are the ones who make Enzo Ferrari his first takeover bid in the early 1960s. It is rumoured to be between 20 and 25 million US dollars and therefore higher than what Ford offered some time later and Fiat ultimately paid.

This one-off from the Nationales Automuseum has a unique history. It has survived the decades in exceptional condition and has been on show at numerous events around the world

Over the decades, #2821GT can be seen again and again at high-ranking events: For example, the special exhibition celebrating Ferrari's 50th anniversary at the Petersen Automotive Museum in Los Angeles, as well as several showings at the Concours d'Élégance in Pebble Beach. By taking part in Concours events, the design icon is helping to cultivate yet another tradition. In addition to the major motor shows, high-society Concours events have offered such exquisite automobiles further opportunities to strut their stuff. There, coachbuilders and designers alike are able to show off their latest designs and use them to promote automotive trends. The most prominent example of such events is the Concorso d'Eleganza Villa d'Este, organised between 1929 and 1952 in the gardens of the palace of the same name on Lake Como.

The two crossed flags with the Ferrari and Pininfarina logos adorn the ashtray. It is located between the seats, behind the gear lever on the centre tunnel

The kink in the C-pillar – from the pen of Pininfarina – long before BM claimed it for itself and declared it the »Hofmeister kink«

The three-eared centre lock on the Borrani spoked wheels is a beautifully designed detail

The logo survives the decades without a facelift

The free-standing headlights are standard on the 250 GT Coupé Pininfarina, white indicators were standard in Italy at the time

FERRARI 250 GT/L »LUSSO«

When Ferrari presented the Berlinetta Pininfarina 250 GT/L in Paris in October 1962, the company showed what it had learnt with the 250 GTO, now transferred to a road-going production model. The »Lusso« marks the pinnacle of the three-litre production models from Maranello.

Pininfarina succeeded in creating an elegant, no-frills line without any decorative elements. Harmonious curves blend with the sparing use of chrome trim. Shortly after the start of production, fog lights were included as standard. The special highlight is the light roof structure with filigree A and C pillars and large, flat sloping windows.

A characteristic element of the equally elegantly designed interior are the large instruments for revs and speed positioned in the centre of the dashboard. Behind the wooden steering wheel, with aluminium spokes, the driver can see five further, somewhat smaller round instruments.

The engine delivers 240 to 250 hp in standard trim. To gain a little more space, it is positioned slightly further forward in the frame than in the racing cars. The chassis, with Watt's linkage, is a direct transplant from the GTO. The limited comfort is also taken directly from the racing sports car. Only around 80 vehicles were produced in just two years. Hollywood star Steve McQueen was one of the customers.

In addition to the motor shows in Turin and Geneva, Ferrari has always favoured Paris as a stage for premiering new models. The motor show was first held in 1898, making it the world's first motor show of international importance. In addition to the good location (in the centre of Europe) and the importance of the French market, the timing (early October) is also an advantage: many car manufacturers launch their new models after the summer holidays.

FERRARI 500 SUPERFAST

The Salon international de l'auto de Genève opened its doors for the first time in 1905. After many years in the centrally located Palais des expositions, the show moves to the Palexpo site in 1982, which is conveniently located near the motorway and airport. The Geneva Motor Show has always marked the start of the European automotive spring and is appreciated by exhibitors for its short distances and clear layout. The traditionally well-represented designers, coachbuilders and small series manufacturers also provide variety.

Ferrari and Pininfarina also regularly present themselves here to the affluent clientele. For example, in 1964 with the premiere of the Ferrari 500 Superfast, the flagship model from Maranello at the time. The twelve-cylinder engine had a displacement of five litres, produced 400 hp and made the 2+2 up to 280 km/h fast. Of course, this also had its price: at DM 99,000, it was twice as expensive as the (also not exactly cheap) Ferrari 275 GTS with 260 hp.

The 500 Superfast is the successor to the 400 SA. With a length of over 4.8 metres, it is also visually imposing. Mockers at the time labelled it the »Royale« of Ferrari. A total of 36 vehicles were built. Bugatti, on the other hand, built just six chassis between 1926 and 1933 for the imposing T41, known as the »Royale«, with a displacement of 12.8 litres and 300 PS.

The 500 Superfast was primarily intended for the US market. However, the first vehicle was delivered by Ferrari importer Auto Becker in Germany.

FERRARI 348 TB

When Ferrari presented the 348, the completely new successor to the successful 308/328 model series, it was quickly nicknamed the »little Testarossa«. In fact, the smooth body surfaces, the side air intakes with their horizontal bars and the rear with the lights hidden behind horizontal struts also suggest such a comparison.

In terms of technology and quality, however, Ferrari's road cars were not yet where Luca Montezemolo would take the company with the successor F355 presented in 1994 – with an aerodynamic underbody, five-valve cylinder heads and 380 hp. And that was a necessary evolution because the 300 hp of the 3.4-litre V8 struggled with the almost 1.5-tonne 348 tb. »auto, motor und sport« editor Wolfgang König writes: "...an average person will drive faster in a GTI than in the 348...«. But then there is also talk of charm, character and acoustic pleasure: »I don't just drive fast, I feel that way too – better than the other way round.«

Ferrari chooses the International Motor Show, or IAA for short, to present the 348. Since 1951, it has been organised every two years in September by the VDA (German Association of the Automotive Industry) at the Frankfurt Exhibition Centre. Ferrari is listed as an exhibitor here for the first time in 1957 – with a small stand in Hall 1, where the German general importer Auto Becker shows three vehicles (two 250 GTs and a 410 Superamerica) to the 800,000 trade fair visitors. There are only two on the 1989 stand: a 348 tb (Trasversale Berlinetta) and the open-top 348 ts (Trasversale Spider). At this point, the new Ferrari has already been sold out for six years.

In the IAA month of September 2003, car orders in Germany rise by 16 per cent – clear evidence of the effectiveness of the trade fair.

Today, the Ferrari factory site in Maranello covers almost 24 hectares. In addition to the racing department and wind tunnel, it also includes modern development and production facilities

»The emotions for Ferrari are not limited to the fans waving flags and wearing caps,« explains Ferrari Formula 1 driver Gerhard Berger. »It is, especially so among the Itali-ans, a quieter love albeit with nuances,

THE DREAM FACTORY

... as if Ferrari were a general term for ›feeling‹, ›love‹, ›passion‹, ›enthusiasm‹, and when you tap into it, something beautiful starts to vibrate, and that has nothing at all to do with cars or racing.«

What began in a workshop with a handful of employees grew into a high-tech factory in which 1802 production employees put 13,221 vehicles on the wheels in 2022. Production figures rise from three units in 1947 to between 30 and 248 in the 1950s. By 1966, the figure was already 928, rising to 2221 vehicles per year in the 1970s. In 1998, more than 4000 Ferraris were sold for the first time, in 2005 more than 5000. 2019 saw the 10,000 mark broken. But regardless of whether just a few models or a broad model range – automotive dreams are always born in Maranello. And since 2021, the world's most important luxury brand has been working in two shifts to fulfil these dreams for its customers.

Ferrari has relied on a high level of vertical integration right from the start. Enzo Ferrari is proud of his highly qualified employees. In addition to the design department and work-shop, he also set up his own foundry early on so that Ferrari could also produce engines and gearboxes entirely in-house. With entry and prize money, the sale of his sports and racing cars and sponsors, Enzo Ferrari juggles his company through the economic waters, no matter how choppy they are. »As far as my company is concerned, I have been asked again and again for years how it was possible for me to take part in so many races and still close the year without a deficit,« writes Enzo Ferrari in his early memoirs. »In the past, races did not require such a massive investment of technical and financial resources. The whole organisation of the races was also not as expensive as it is today. With the help of the car accessory companies, which were able to capitalise on the publicity, it was sometimes possible to make a profit.«

But the support of Fiat is also important. Again and again. In the »auto, motor und sport« interview conducted by Giovanni Lurani in 1958, Enzo Ferrari said: »I have always taken the view that victories, especially abroad, are good for the reputation of the entire industry in the country concerned.« Fiat also sees it that way. For the 1956 Formula 1 season, for example, the Turin-based car giant steered the unrivalled D50 monoposti from the economically ailing company Lancia to Ferrari. In the mid-1960s, Fiat provided support with the Dino sports model project, which ensured respectable production figures, overcame a homologation hurdle and expanded the Ferrari model range. In 1969, Enzo Ferrari finally sold 50 per cent of his company shares to the Turin-based group – with the agreement that the majority of the remaining shares would also go to Fiat after Enzo's death.

As part of the sale, with which Ferrari regulated the succession and continued existence of the company he had founded after the death of his son, Fiat sent two managing directors for the sports car business to Maranello. Francesco Bellicardi (formerly Weber S.p.A.) as well as Carlo Pelloni, Oscar Montabone (Fiat), Ermanno della Casa (Ferrari), Sergio Pininfarina and Enzo's son Piero Lardi sit on the Board of Directors. »Now that the people at Fiat were taking care of the daily routine of designing, building and selling road cars, Ferrari was relieved of a task that had never interested him. For him, customer cars had always been a means to an end, and the older he got, the more he despised

The very first Ferrari rolled through this gate. Here, Enzo Ferrari is accompanied, most likely to the Cavallino restaurant opposite, in 1977

The classic gateway to the old Ferrari factory has long been located in the heart of Maranello. It is still an unchanged part of the daily hustle and bustle

This is what production looked like in the early years of the company. Here you can see work on the Grand Prix racers and sports cars of the early 1950s. The pit is also worth noting

During the final preparations for the 1957 Mille Miglia, the sports cars gather in the factory courtyard

the lavish nouveau riche and the empty-headed aristocrats…«, writes Ferrari chronicler Brock Yates. Fiat invests millions in the factory, the production facilities and the modernisation of the foundry. By 1970, the workforce already numbered 1000 employees. The new Fiorano test track is added. It replaces the track in Modena and cumbersome trips to Monza. With integrated television cameras and telemetry installations, it becomes the most modern test track in the world. After the first (and until then only) plant expansion in 1962, the reorganisation in 1970 marked an important milestone in the production history of the sports car manufacturer. Also in the 1970s, Scaglietti was gradually integrated into the company. Further milestones followed in 1978 and then, in increasingly rapid succession, in 1981, 1982, 1985…

After the death of Enzo Ferrari on 14 August 1988, Fiat held 89 percent of the shares, 10 percent remained in the hands of Piero Ferrari and one percent was owned by Pininfarina. After Fiorano, Ferrari expands its testing facilities in 1988 with the purchase of the circuit in Mugello.

The modern production line in Maranello is reminiscent of a clinic or Formula 1 pit. The GTs have already had their heyday here

In 1991, Ferrari is in financial difficulties. The speculative bubble with exotic sports cars – above all Ferrari – bursts. As a result, the usual price premiums are no more. The plant has to halve its production. And to be on the safe side, Fiat separates the brand from the company and incorporates it into its private holding company in Luxembourg. A new head of the company is also appointed. Luca Cordero di Montezemolo, an old acquaintance, returns to Maranello. With international management experience under his belt, the company is reorganised under his leadership – and thanks to financial injections from Turin.

In terms of road-going sports cars, he pushes quality and excellence with new launches such as the F355 and the 550 Maranello. And he also set the course for a renaissance of the Reds on the racetrack. For the 1992 season, Montezemolo engaged his old companion, airline entrepreneur Niki Lauda, as a consultant. With the other personnel decisions that resulted in the »Dream Team«, Montezemolo and Ferrari scribe lasting history.

In 1947, the company premises covered an area of 10,000 square metres; in 2006, they extended to 131,461 square metres. Added to this are the 15,467 m^2 of the Scaglietti bodywork plant in Modena, where the aluminium bodies for the GT cars are produced. While Enzo Ferrari's team comprised 250 people at the beginning of the 1950s, the entire Ferrari workforce numbered 4919 in 2022.

Return of the successful Daytona participants in 1967 – the third, not depicted 330 is now part of the Loh Collection

A look into the racing car workshop during the 1950s: amateur racing driver Tony Parravano visits the production line. He was one of the regular customers at the time

»The moment will come when I will see my cars doing their rounds on Sundays, simultaneously in two or three different parts of the world. And they will win two or three times on the same day,«

OFF YOU GO!

... said Enzo Ferrari to his journalist friend Gino Rancati in 1944, while he was trying to protect his machine tool factory from bombs and looters - at least that's the story Rancati tells.d. Five years have passed since Enzo Ferrari's separation from Alfa Romeo, and the non-compete clause in his termination agreement is coming to an end. And yet – due to the war – it was to be another three years before he presented the first car bearing his name. But Enzo Ferrari was proved right: His racing cars would go on to win Grand Prix, hill climb and sports car races. Over and over again.

»Apart from the fact that, in my opinion, there is no ideal racing car, as a designer I am interested in every new formula – provided that it is not so simplified that it makes excessive financial demands that are unrealisable for me,« explained Enzo Ferrari to editor Giovanni Lurani in 1958 for an interview published in »auto, motor und sport«. »I am convinced of the technical value of most races, as they represent an incomparable and irreplaceable test.«

February, the 24th, 1962 was a day that clearly demonstrated the range: On this sunny winter's day, Enzo Ferrari holds a press conference in Maranello with the racing cars for the upcoming season. The date is now well known, as the media representatives get to see the 250 GTO for the first time. More important at the time is the latest version of the 156 F1, with which the Scuderia intends to defend the world championship title won by Phil Hill.

Those present will also be able to see four new mid-engined racing sports cars with identical bodywork. They go by the names 196 SP, 246 SP, 286 SP and 248 SP. Depending on the race track and regulations, either a Dino six-cylinder with 1984, 2417, the new Chiti 90° V6 with 2862 cc displacement or the new 90° V8 with 2458 cc, also designed by Chiti, is bolted into the tubular frame. The chassis construction is identical to that of the 246 SP (Sport Posteriore) shown the previous year – now with the F1-proven mid-engine design.

A new feature was the installation of a flatter windscreen, which was permitted from 1962. Accordingly, the featherweight aluminium body, based on wind tunnel tests, was also flatter. Only for the 24 Hours of Le Mans, which was held in accordance with ACO regulations, did the all-round Plexiglas windscreen have to be extended by a higher glass element.

However, when the new sports cars were presented to the assembled press, their creators were no longer involved. The engineers Carlo Chiti and Giotto Bizzarrini, who joined Ferrari from Alfa Romeo in 1957, successfully convinced Enzo Ferrari of the mid-engine design, the use of disc brakes, the abandonment of spoke wheels and showed him the importance of aerodynamics. These and other ideas became the key to the renewed strength of Scuderia Ferrari. The young Mauro Forghieri, who had joined Ferrari from university in 1959, supported the two veterans to the best of his ability. Despite all the successes in the 1961 season, there was still discord within the company, which the Ferraristi later labelled the »pal-

In February 1961, the new Formula 1 Ferrari 156 and the six-cylinder 246 SP mid-engined sports car stand side by side in harmony

The two-part radiator opening became a characteristic design element for the racing cars from Maranello at the beginning of the 1960s

In the course of its service life, #0790 underwent a series of modifications. These included the revision of its aerodynamics with a flatter body

ace revolt«. Whether this had to do with Enzo Ferrari's wife Laura becoming increasingly involved in the fortunes of the racing department, or whether it had to do with upheavals in the ever-growing company, can no longer be determined today. In any case, Enzo Ferrari kept a firm grip on the reins when he clashed with commercial director and head of sales Girolamo Gardini. Together with Gardini, a large part of the management team left Maranello at the end of 1961: head of development Carlo Chiti, Giotto Bizzarrini (head of the prototype department), foundry manager Fausto Galassi, personnel director Enzo Selmi and Scuderia Ferrari racing director Romolo Tavoni also left. Only the young Mauro Forghieri remains on board – and is immediately promoted to Sports Director and Chief Designer.

Immediately after their departure, the renegade sextet founded the sports car manufacturer ATS (Automobili Turismo e Sport). This annoyed Ferrari – but not for very long, as ATS remained an unsuccessful venture. Bizzarrini subsequently made a name for itself with various sports car projects, and Carlo Chiti founded Autodelta, the new

The heart of the racing car beats with six cylinders and comes from Dino Ferrari, at least officially

The streamlined rear-view mirror bears the marks of decades with dignity

racing division of Alfa Romeo. Under his aegis, the Giulia TZ and GTA models and the Tipo 33 were created, and Alfa Romeo won numerous international and national championship titles with the cars prepared by Autodelta. It then found its way into Formula 1 from 1977.

The 246 SP on display at the Nationales Automuseum has chassis number #0790. It is the first representative of this model series and is presented a year earlier at the annual Ferrari press conference – originally still with a high windscreen and voluminous rear body. In its first race, the 12 Hours of Sebring at the end of March 1961, Wolfgang von Trips and Richie Ginther led for hours before retiring after a spin with a bent front axle. The 246 SP then proved its potential in the exciting Targa Florio held in April. Gendebien, von Trips and Ginther share the wheel. They were able to hold their own against both their teammates and the fastest Porsche team. Three weeks later, Wolfgang von Trips also secured the first mid-engine victory for Ferrari in Formula 1 at Zandvoort, while von Trips/Ginther/Gendebien took the chequered flag in third place in the icy cold 1000-km race at the Nürburgring.

The vehicles are built by the Modena-based Carrozzeria Fantuzzi. Below the front you can see one of the hooks to which the high-speed jacks are attached

As befits a racing car: the rev counter dominates the instruments

Today, drivers adjust numerous parameters using rotary controls on the steering wheel. Back then, the pilot reached for the stopcock

The striking front is typical of the models from a time characterised by a »palace revolt« at Ferrari

At that time, the bonnets were still secured with leather straps to prevent them from popping open unintentionally

The engine number belongs to the chassis, which has been fitted with different engines over the years

And at Le Mans, Ginther/von Trips set the fastest race lap with #0790, but retired at eight o'clock on Sunday morning due to incorrectly calculated fuel consumption.

In 1962, the 246 SP was then given the flat body described above by Carrozzeria Fantuzzi. Fantuzzi was another coach-building company based in Modena since 1939, which Ferrari used from time to time to avoid production bottlenecks. Fantuzzi was originally associated with Maserati. However, after Maserati stopped its factory racing activities at the end of the 1950s, the brothers Medardo and Gino Fantuzzi increasingly coachbuilt Ferrari's racing cars – right up to the 1964 1512 F1 also shown in the Loh Collection. Incidentally, Fantuzzi remained in the business of restoring historic racing cars until 1993.

There was also a rich harvest for #0790 in 1962: Olivier Gendebien and Phil Hill won at the Nürburgring. In Sebring (Pedro and Ricardo Rodriguez) and Mosport (Innes Ireland/Mike Parkes), retirements after leading by kilometres are etched in memory. But at the end of the season, there was another victory at Brands Hatch (Mike Parkes). The 246 SP #0790 was then converted to the 196 SP for the 1963 season. In this configuration, Ludovico Scarfiotti dominated the 1962 European Hillclimb Championship with a sister car. He won four of the seven rounds of the championship and came second once. As only the five best

Behind the compact V6, you can see the longitudinally mounted 5-speed gearbox with clutch bell housing. The starter motor is positioned above it. By the way, treadless slick tyres had not yet been invented at that time

results counted in the European Hill Climb Championship, he collected 42 points in the points system adopted from F1. Porsche drivers Walter and Greger, who were chasing him, only scored 33 and 30 points respectively.

Scarfiotti becomes Italian Hillclimb Champion several times. He goes by the nicknames »Dodo« and »Lulu« and is by far one of the most popular Italian drivers in the field. Ludovico Scarfiotti is the grandson of his grandfather of the same name, who founded the Fabbrica Italiana Automobili Torino FIAT together with Giovanni Agnelli in 1899. Scarfiotti's father is an engineer and also races cars. After winning the European Hillclimb Championship, Ferrari entered him in other sports car races. He achieved major victories at Sebring (1963) and Le Mans (1963 and 1964) as well as at the Nürburgring (1964). In 1965, he won the European Hillclimb Championship title once again, this time in a Ferrari Dino 206 P. He was also allowed behind the wheel of an F1 Ferrari six times between 1963 and 1967. He won the home race in Monza in 1966. For the 1968 season, he switched to Porsche and into the cockpit of the ultra-light Porsche 910 Bergspyder. He loses his life in it during training for the second championship race at Roßfeld.

For #0790, its career as a 196 SP continued on hilly terrain: Edoardo Lualdi-Gabardi took to the wheel; a gentleman driver who primarily contested mountain rac-

Set aside after a successful mission – the 246 SP/196 SP with VIN #0790 on display at the Nationales Automuseum is also fascinating because of its unique patina

Registration the easy way: At the time, road-legalisation was still part of the sports car regulations. In Maranello, the vehicles were given their licence plates on the spot with a brushstroke

es until 1972. He began the season with the two-litre racing sports car, and team colleague Umberto Bini, with a start at the Targa Florio. Unfortunately on Sicily the car broke down on the first lap. However, he went on to win the subsequent two races on Italian soil in the mountains. In the following competitions at Mount Ventoux, Trento-Bondone, Cesana-Sestriere and Ollon-Villars, the accomplished amateur narrowly missed out on the podium.

In January 1964, #0790 was then retired by the works team. Ferrari racing director Eugenio Dragoni signs a purchase agreement with the Milan-based Scuderia Sant'Ambroeus. Dragoni had once co-founded the private racing team in the 1950s. After the surprising departure of Romolo Tavoni, he became the racing director of Scuderia Ferrari. Today, the vehicle on display in the Nationales Automuseum impresses with its carefully preserved patina, which tells the story of an eventful racing career.

PORSCHE 910 SPYDER, BERGSPYDER

In 1967, Porsche decided to only enter factory-new cars in important races. This meant that improvements could be introduced and new models developed in quick succession. The 906 »Carrera 6« had barely hit the track when the 910 appeared, with which Porsche took lightweight construction to the extreme. By using materials such as aluminium, magnesium, titanium and beryllium, an alloy tubular frame construction and paper-thin fibreglass bodywork, Porsche engineers achieved a total vehicle weight of 382 kilograms at the end of 1968 – for a ready-to-drive racing sports car with a very heavy eight-cylinder engine, mind you!

Lightweight construction was also the maxim for the sister model built for endurance racing. The coupé, built with both a six and an eight-cylinder engine, has an aerodynamic body with a removable roof. The small 13-inch wheels are made of magnesium. Another new feature is the central locking system, which allows the individual wheel nuts to be quickly fitted and removed using a pneumatic hub-nut driver.

As Porsche had recently given its customers the 906 and did not want to offer a new model for sale in the following season, the 910 was originally only used by the factory and in the prototype category. Porsche builds a total of 28 cars, each of which is only raced once. Afterwards, they are given a general overhaul and end up in the hands of customers.

1967

ALFA ROMEO TIPO 33/2 »DAYTONA«

For the debut of the Tipo 33, Alfa Romeo sensibly chose a hill climb; the race in Fléron, Belgium. There, the Autodelta works team, presented a new two-litre racing sports car. It is nicknamed the »Periscopa« because of its air scoop above the roll bar. With the new 90° V8 developed under the direction of ex-Ferrari manager Carlo Chiti, the mid-engined car weighs 580 kilos ready to race.

Despite the victory at its premiere, the »Periscopa« was only granted a short summer. The delicate suspension components attached to the magnesium front subframe proved to be the Achilles' heel of the vehicle, which was otherwise superior to the competing Porsche. In February 1968, the Tipo 33, with its repositioned radiators and more angular fibreglass bodywork, made a big impression at the Daytona 24-hour race: Now the cars are not only fast, but also reliable. They finish in fifth, sixth and seventh place. Alfa Romeo is happy to recreate a group photo à la Ferrari in 1967 as the cars cross the finish line. From then on, the model was given the type designation »Daytona«, even though Porsche crossed the finish line in the first three places in the same class – also prompting a group photo.

At the Targa Florio, the audience witnessed four Daytonas in the top six, including a class victory in the prototypes with a displacement of up to two litres. At the 1000-kilometre race at the Nürburgring, another class victory is on the cards, as well as at Le Mans, where Giunti and Galli only just miss out on the podium in fourth place overall. The other 33s finished the race directly behind them and ensured another triple triumph in their category. At the end of the season, Alfa Romeo finished third in the manufacturers' world championship, behind Ford and Porsche. The following year saw the debut of the 33/3.

1968

ABARTH 2000 SE010

The Viennese Karl alias »Carlo« Abarth came to motorsport as a sportsman via bicycles and motorbikes. After the war, he started as sports director of the soon insolvent manufacturer Cisitalia. From the remnants of Cisitalia, the Italian by choice began his career as a manufacturer of exhaust systems, tuning parts and racing cars. For a long time, Abarth designs dominated races of all kinds, albeit in the smaller cubic capacity classes.

In the second half of the 1960s, Carlo Abarth also added a new two-litre model to the 1000cc models. In 1969, for example, Arturo Merzario won the sports car classification of the European Hill Climb Championship with the Abarth 2000 SE010 Sport Spider. Abarth opted for a rear-mounted engine due to the favourable weight distribution for driving uphill. This produces 250 hp.

While the predecessor model still competed in the racing car classification, the officials categorised the Abarth Sport Spider as a sports car because of its looks. This had the disadvantage that Abarth had to prove that it had built a total of 25 examples of this vehicle. Fortunately, Abarth has long standing reputation and therefore a willing customer base. The classification also has the advantage that Arturo Merzario does not have to compete in the same category as Peter Schetty.

The Swiss racing driver switched from Abarth to the Ferrari camp for the 1969 season and drove the 290 hp Ferrari 212 E Montana prototype. He was, however, no match for it in the 1969 season. Oftentimes, Schetty took 20 or even 30 seconds off Merzario. But in the end, the Italian did manage an overall victory. And with four second and one third place, it was a remarkable end to the season for Abarth and Merzario.

1968

Racing has experienced long-distance races from the very beginning. Ideal for demonstrating performance and reliability. In Germany, the Nürburgring 1000 km race marks the highlight of a sports car year

RED GLAMOUR IN THE GREEN HELL

Until 1950, Germany was excluded from international motorsport competitions and racing was out of the question on the war-torn roads. Even the Nürburgring, which opened in 1927, first had to be repaired. However, with the announcement of a marque world championship in 1953, the traditional circuit became the focus of the FIA, whose world championship allowed one circuit per nation. And as far as the race distance was concerned, the German organisers were happy with 1000 kilometres (based on the Mille Miglia); this set the framework. Ferrari sent a prominent car to the first 1000 km race in the Eifel in 1953: at the wheel were Formula 1 world champions Alberto Ascari and Nino Farina, who won in their 375 MM after the retirement of the favoured Lancia.

It looks like love at first sight. But in fact, both the organiser and Ferrari needed a little patience in the time that followed. For the ADAC, the premiere turned out to be a loss-making event due to the mere 8,000 to 10,000 spectators. In 1954, it cancelled the race when the »driving force« Mercedes did not take part, and in 1955 the event was cancelled in the wake of the Le Mans disaster. In 1956, 70,000 spectators watched a duel between Ferrari and Maserati, with the Reds from Modena coming out on top. The Maserati 300 S of Taruffi/Schell/Behra/Moss finished ahead of the Ferrari 860 Monza of Fangio/Castellotti and the Ferrari 290 MM, which was shared by Hill, Wharton, de Portago and Gendebien.

Here Carlos Pace boards the carousel, a section of the Nürburgring-Nordschleife that remains legendary to this day, in his 312 PB. He finishes second in the 1000-kilometre race in 1973

In the three years that followed, victories went to Aston Martin. Although Ferrari won the World Sports Car Championship title in succession, the Scuderia had to make do with podium finishes at the Nürburgring. The day-long race in the »Green Hell« (as Jackie Stewart later referred to the circuit) had long since developed into a crowd-puller. Between 150,000 (1959) and 400,000 (1965) spectators came to the Ring.

In 1960, Ferrari hit the headlines when a fire broke out in the pits. It ignites as the Dino 246 S of Scarlatti/Cabianca/Seidel readies for refueling. The flames not only engulf the vehicle, but also the garage and the petrol tank on the pit roof. After two Maserati victories (1960 and 1961), 1962 was finally Ferrari's finest hour. Phil Hill and Olivier Gendebien won with a Dino 246 SP ahead of their teammates Willy Mairesse and Mike Parkes, who had started in a 250 GTO. Wolfgang Seidel and Peter Nöcker finished sixth in Seidel's 250 GT SWB. The 246 SP won the sports car classification, the 250 GTO won the prototype class and the 250 GT came first in the GT category. The 1000 km race in 1962 marked the start of a Ferrari winning streak that lasted until 1965.

The pack of mid-engined sports cars from Maranello chases through the »Green Hell« in unison. Nino Vaccarella leads ahead of Lorenzo Bandini and Innes Ireland. The victorious 275 P can be seen today in the Nationales Automuseum

Mike Hawthorn leans against his 1957 315 S parked outside the pits and chats with Ferrari team manager Romolo Tavoni. Hawthorn/Trintignant finish third, their team-mates Collins/Gendebien cross the finish line one and a half minutes ahead of them in a 335 S

In 1963, there was a one-two-three to celebrate. John Surtees and Willy Mairesse won with the 250 P entered by the factory, ahead of the private 250 GTO of Noblet/Guichet and the 250 TR 61 (Abate/Maglioli) entered by the Scuderia Serenissima. The following year, Ford celebrated its Ring première with the GT40 and Ferrari suffered a number of retirements. But in the end, the 275 P of Scarfiotti/Vaccarella won ahead of Parkes/Guichet and their 250 GTO/64. And in 1965, the Ferrari 330 P2 of Surtees/Scarfiotti crossed the finish line 45 seconds ahead of the second-placed 275 P2 (Parkes/Guichet) after almost seven hours of racing. Bandini/Vaccarella were classified fourth in their Ferrari Dino 166 P.

After that, the Nürburgring years dominated by Porsche began. Ferrari only sporadically entered a few cars into the 1000-kilometre race. For example in 1970, when Surtees/Vaccarella finished third with the 512 S. With the 312 PB, Ferrari once again had a very hot iron in the fire. And the race, which is still part of the World Sports Car Championship, is also on the Scuderia's travelling calendar. In 1972, the Ferrari teams Peterson/Schenken and Merzario/Redman won ahead of the Alfa Romeo Tipo 33 of Marko/de Adamich/Hezemans. And in 1973 there was another Ferrari one-two with Ickx/Redman and Pace/Merzario. However, only 30,000 spectators came to the Ring in 1972. At the end of the 1973 season, Ferrari finally withdrew from international sports car racing. From then on, the 1000 km race became a playground for Group 4 and Group 5 cars, before the Group C sports cars once again reached a high point. The end of the Group C cars also marked the end of the continuous history of the Eifel endurance classic. The races are then only held irregularly (2000, 2004-2007 and 2016/2017). In 2004, the 550 Maranello, entered by the Larbre Compétition team, achieved a victory for Ferrari in the GTS category at the Grand Prix course, organised as part of the Le Mans Endurance Series.

The Marathon de la Route, held between 1965 and 1971, was another endurance racing episode. The race, which lasted 82 or even 96 hours, emerged from the Liège-Sofia-Liège road race, which was moved to the Nürburgring for traffic and safety reasons. Incidentally, Luca Montezemolo also competed here in 1969. He finished ninth in a Fiat 125 S. 1970 marked the beginning of the 24-hour race's history, which continues to this day and remains loyal to the Nordschleife even after the opening of the Grand Prix circuit (1985). Every year, up to 200,000 spectators come to watch the field of 180 cars on the traditional Eifel circuit. In 2023, they witness a Ferrari victory for the first time: the Bamber/Catsburg/Pittard/Fernández Laser team triumphs in the Ferrari 296 GT3 entered by the Frikadelli team.

24 HOURS OF LE MANS

The endurance classic on the Sarthe is probably the most famous car race in the world – rivalled only by the Monaco Grand Prix and the 500 miles of Indianapolis. Since 1923, the race has become a fixed star in the motorsport calendar. Ferrari won the 24-hour race ten times between 1949 and 2023. In 1954, Trintignant/Gonzáles followed up the premiere success of Chinetti/»Lord Selsdon« (166 MM) in the first post-war race in 1949 with a 375 Plus. Major victories followed in the 1950s and 60s with various versions of the front-engined Testa Rossa racing car (1958, 1960, 1961 and 1962) and the mid-engined 250 P (1963), 275 P (1964) and 250 LM (1965) models. In 1973, the factory competed in Le Mans for the last time for a long while. However, the 312 PB, which started from pole position, had to settle for second place at the finish, six laps down. The 365 GTB/4 »Daytona« Competizione, which was also part of the Loh Collection, repeated its class victory that year.

After that, another 50 years would pass before Maranello would once again fight for overall victory. In 2023, the new 499 P Hybrid achieved a real feat: victory for Guidi/Calado/Giovinazzi, fifth place for team-mates Fuoco/Molina/Nielsen.

In the meantime, however, Ferrari has also become one of the regular protagonists on the 13.626-kilometre high-speed circuit. Specifically, the racing versions of numerous twelve- and eight-cylinder models (such as the 512 LM, 348 LM, F40 LM, 360 Modena, F430 GT2, 458, 488 GTE) used by various teams, as well as the 333 SP prototype initiated by Piero Ferrari and Giampietro Moretti. In total, the Ferrari teams scored a further 36 class victories at Le Mans, most of them without factory support, some of them with minimal factory support.

24 HOURS OF DAYTONA

The 24 Hours of Daytona was launched in the mid-1960s when the automotive Goliath (Ford) fought a duel with the sports car David (Ferrari). Sports cars have been competing on the track on the east coast of Florida since 1959. Daytona has been part of the World Sports Car Championship calendar since 1962, initially as a 3-hour race and from 1966 onwards with a duration of 24 hours. For the Americans, the endurance test held on parts of the race track oval as well as a winding track inside the circuit serves as a dress rehearsal for the classic at Le Mans.

The premiere event goes to Ford – although Maranello is »skipping« this World Sportscar Championship race. The Ferrari cars represented are entered by Luigi Chinetti's North American Racing Team (NART) and other privateers. In 1967, a success was achieved with a commanding one-two-three victory at the Detroit car giant's home race, which motivated Ferrari to decorate the Berlinetta 365 GTB/4 presented in Geneva in the spring of 1968 with the model designation »Daytona«. One of the cars from the winning trio, the 412 P by Rodriguez/Guichet entered by NART, is also part of the Ferrari exhibition at the Nationales Automuseum.

The 24-hour race, which is always held at the beginning of the season, has been part of various championships – such as IMSA, USA Road Racing and the Grand-Am Sports Car Series. After the sports car prototypes, the Group 5 and IMSA GT/GTP racers, or Group C and IMSA GTP cars are entered – categories that Ferrari itself chooses not to enter.

In 1999, visitors to the Daytona 24 saw red again: Theys/Baldi/Luyendyk/Moretti won in the Ferrari 333 SP entered by the Doran-Moretti Racing Team.

24 HOURS OF SPA-FRANCORCHAMPS

In 1924, just one year after the 24-hour race premiere at Le Mans, the Royal Automobile Club of Belgium put together a starting field on a 15-kilometre circuit over the country roads between the Ardennes villages of Francorchamps, Malmet and Stavelot, marking the beginning of the history of the Spa-Francorchamps endurance classic.

In the 1930s, Scuderia Ferrari with drivers Brivio/Siena, Chiron/Chinetti and Sommer/Severi won three times in a row. The oxblood red, supercharged Alfa Romeo 8C 2300 and 8C 2900 A cars proved to be unbeatable. The latter also won once again in 1938 as part of a works race, for which Enzo Ferrari was responsible.

After the war, Luigi Chinetti also triumphed at the 24 Hours of Spa-Francorchamps following his victory at Le Mans. Once again, the French and American Ferrari importer was at the wheel of a Ferrari 166 MM, which he shared with Frenchman Lucas. The next twice-around-the-clock race was not held until 1953. This time, Nino Farina and Mike Hawthorn – with a 375 MM – repeated the success for Ferrari.

When the history of this race continued after another break in 1964, the touring cars and Grand Tourisme took part. The endurance race usually counts as a round of the European Touring Car Championship. For top-class sports cars, the 1000-kilometre race is held in Spa from then on. But sports cars also return to the 24-hour race – mainly Porsche 911s, which also characterise the starting fields of the 24-hour race at the Nürburgring. In the Eifel, the sports cars from Maranello remain a rare side note – until victory in the 2023 edition (Bamber/Catsburg/Pittard/Fernandez Laser in a Ferrari 296 GT3).

Ferrari is often conservative. Accordingly, Maranello was late to adopt the long-established mid-engine design. But when the race cars with the engine behind the driver finally started to compete, they became instant winners

SCORING A BULL'S EYE

For Enzo Ferrari, the development and production of road-going sports cars was both an annoying evil and an accepted necessity: he used the money raised to finance his passion for motorsport. And to ensure continued success in various fields – from Formula 1 and Formula 2 to sports car and hill climb racing to the Tasman and CanAm series – Ferrari constantly developed new sports hardware in the 1960s. Always with an eye on the best possible performance in terms of engine power, weight and chassis rigidity. The better is the enemy of the good. New racing cars are launched in quick succession for every season – across all playing fields.

After Ferrari's first Le Mans victory (1949, 166 MM), the sports cars from Maranello won again in 1954 and 1958 with the 375 Plus and 250 TR 58 models. Six consecutive victories followed between 1960 and 1965. The successful racing cars were given the model designations 250 TR 59/60, 250 TRI/61, 330 TRI LM Spyder, 250 P, 275 P and 250 LM/275 LM. The many different vehicle models and engine capacity variants were also due to the fact that FISA (the branch of the FIA responsible for motorsport at the time) was constantly changing the regulations. In the era of increasing mass motorisation, the distinction between GT sports cars and prototypes is not yet set in stone. In addition, some race

Form follows function – the fuel filler cap on the front wing is a functional as well as aesthetic element of this Gran Touring sports car

The 275/330 P with chassis number #0820 is in exceptional condition. Today, it is regarded by Ferrari as a reference vehicle

The rollover protection on the the 275/330 P may well be rudimentary, but it set a worldwide trend on the streets and preceded designs such as the »Targa bar«

In the dark, lamps illuminate the start number

The longitudinally mounted V12 lives under the bonnet

organisers (such as the ACO) are taking alternative paths that sometimes run contrary to FISA's plans. For example, FISA wants to strengthen the category of GT cars derived from production vehicles. Accordingly, it only organises the marque world championship for this vehicle category from 1962 onwards. The ACO, on the other hand, is backing the spectacular sports prototypes that are so popular with the public. In cooperation with the organisers of the Targa Florio, the 1000 km race on the Nürburgring and the 12 Hours of Sebring, the titles of »Sports Car Cup« and »Challenge Mondial de Vitesse et d'Endurance« were created without further ado. In the mid-1960s, the sports car battles at the Sarthe or in the Eifel captivate large crowds: Le Mans 1964 attracts 350,000 spectators, the 1000-km race 300,000.

In 1961, Ferrari presented the six-cylinder 246 SP, both models with a mid-engine, alongside the F1 156 model. In addition to the balanced weight distribution, aero-

The open gearshift gate of the racing car can also be found in the road models. For a long time, it was characteristic of Ferrari vehicles

Scarfiotti, Vaccarella, Parkes, Rodriguez and Piper gripped this exquisite steering wheel

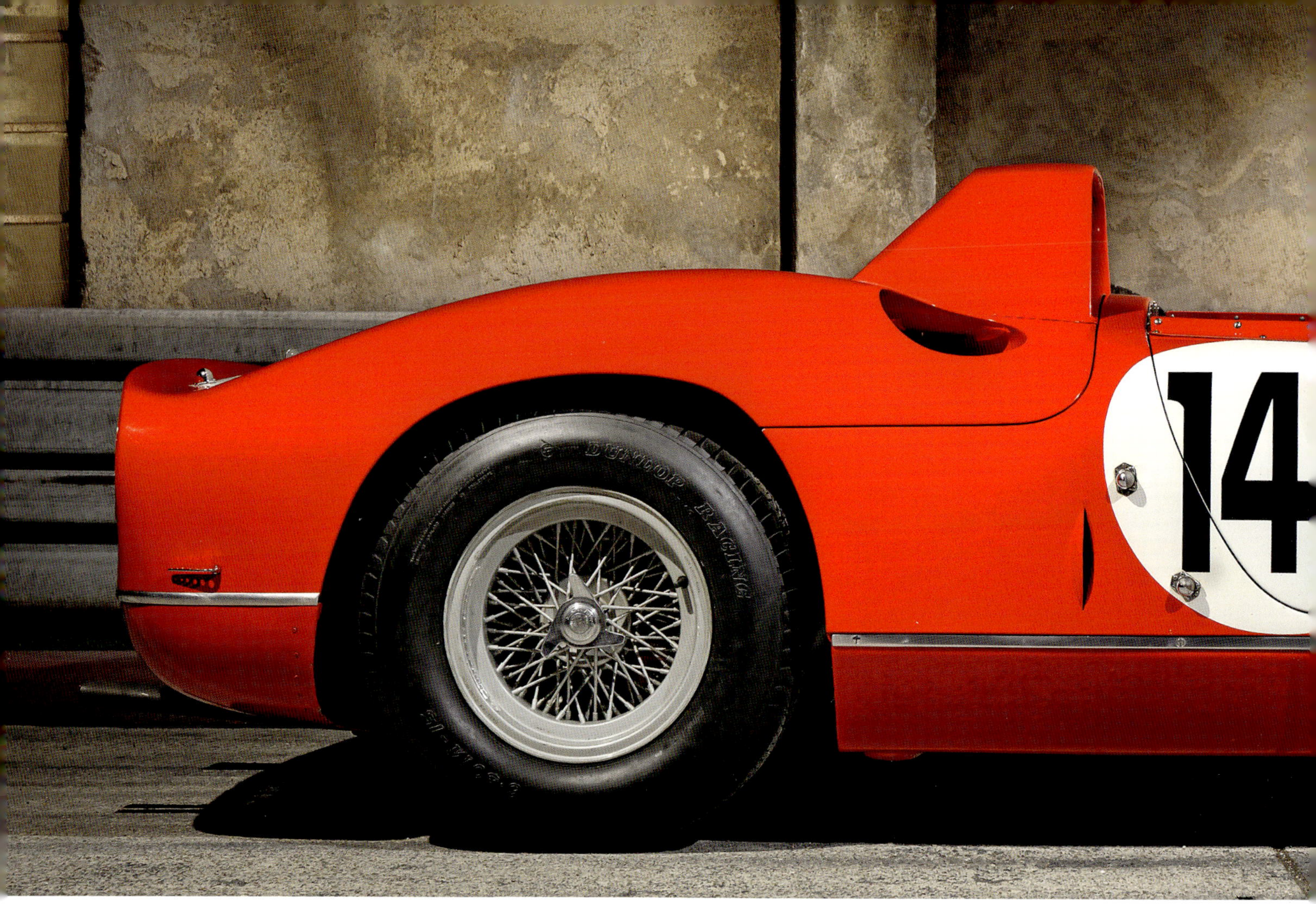

The mid-engine design gives the Ferrari prototypes a new silhouette

The two distributors of the twelve-cylinder engine can be seen in the foreground, with the two ignition coils mounted between them. The V12 has a twin-plug design with two spark plugs per cylinder

The brakes are flange-mounted to the gearbox, which sits behind the engine. Air flows through the intakes to feed the power plant.

The illumination is considered, at most, modest for night-time use. The Plexiglas covers absorb further lumens

dynamic advantages also spoke in favour of this design. However, everything remained the same for the twelve-cylinder sports prototypes until the appearance of the 250 P (1963) – with a further developed Testa Rossa (TR) in the proven front-engine design. The GT sports cars also continued to have the classic vehicle layout for the time being. The P stands – depending on the interpretation – for »Prototipo« or »Posteriore«; the latter refers to the installation location of the engine and is clearly Ferrari's preferred interpretation... Why? The company also used the 250 P as the prototype for the 250 LM presented at the end of 1963. This was to be officially homologated as an evolution of the 250 GT or 250 GTO – as a series-produced GT vehicle, not as a prototype.

A stylistic innovation is the bar behind the cockpit of the 250 P. It is a response to emerging safety discussions and will establish the fashion for »Targa« and roll-over bars for convertibles. For the 1964 season, the 250 P was further developed into the 275 P and 330 P. At that time, Ferrari swapped the 3.3 and 4-litre V12s in a checkered fashion. As a result of the rather lax bookkeeping at the time, it is not possible today to fully trace which engines were used in which chassis, in which race. In addition to the four cars from the previous year (#0810, #0812, #0814, #0816),

Depending on the race and driver line-up, Ferrari opted to use 3.3 or 4-litre engines in this vehicle. Hence the designations 275 P and 330 P respectively

There are two valves per cylinder, and one overhead camshaft per cylinder bank. Three cylinders share one of these four exhaust pipes

The rev limit is pragmatically taped off. RPM, water and oil temperature, oil pressure and fuel level – provide sufficient information for a safe arrival at the destination

which had also been converted to 275 and 330 engines, three more cars were built for the 1964 season (#0818, #0820, #0822) and a fourth for the following year (#0824).

The Nationales Automuseum's 275/330 P bears the chassis and engine number #0820 and is driven by Ludovico Scarfiotti, Nino Vaccarella, Mike Parkes, Pedro Rodriguez and David Piper. Its career begins with a commanding victory for Scarfiotti/Vaccarella in the 1000-km race at the Nürburgring. It is highly likely that a 330 engine is installed in the Eifel. With 370 hp, the 4-litre is more powerful than the 3.3-litre engine, but with a total weight of 785 kg, it also weighs 30 kilos more. This is why some drivers prefer the weaker but more manoeuvrable model. Both versions prove to be a firm favourite. In Nürburgring qualifying, John Surtees, in the sister car, beat the 9-minute barrier by 3 seconds – quite a sensation at the time.

At Le Mans in 1964, no fewer than eleven Ferraris lined up at the start. In addition to a quartet of 250 GTOs and a 275 LM, three 275 P and three 330 P cars started. Mike Parkes and Ludovico Scarfiotti took turns at the wheel of the Loh Collection car. Behind them is a 275 V12. While the race ends for #0820 on lap 71 with oil pressure problems, Jean Guichet and Nino Vaccarella win in the 275 P ahead of the two 330 P cars of Hill/Bonnier and Surtees/Bandini. After the 24-hour marathon, Ferrari sends the 275 P across the Atlantic for the rest of the season. There, #0820 will contest races in the USA, Canada and the Bahamas. At the end of the racing season in Nassau, the car is sold and passes into the hands of the Mecoms. Thanks to the import of oil from the Middle East, they have a rich source of finance. Mecom Sr. and Jr. use their money not only to finance their passion for racing and their Ferrari collection – they even make Enzo Ferrari a takeover bid at the time. However, when Ford entered into negotiations with Ferrari, the Mecoms withdrew their offer. In the hands of the Mecom team, the face of the now used racer changed. It was given a flat, all-round Plexiglas windscreen, got rid of the bar behind the seats and was painted in the US racing colours – white with blue stripes.

After various stops in the hands of US collectors, the car became part of the Obrist Collection in 1986. Ten years later, the then Formula 1 boss Bernie Ecclestone acquired the collection and with it the 275 P-64 with chassis number #0820. After a further stay in the US, this prototype with a complete CV has been part of the Loh Collection since 2008. »Ferrari Classiche« rates it as a reference vehicle for the 275/330 P model. The accident-free car has long since been restored to its original delivery condition using the original parts that have been preserved over the decades.

FERRARI 750 MONZA

In June 1953, Alberto Ascari starts at the Gran Premio dell'Autodromo di Monza with the 735 S, Ferrari's first sports prototype with a four-cylinder engine. The following year, the engine derived from the Formula 1 Ferrari was further developed into a three-litre unit. Instead of 225 hp, the racing car now produced 250 hp and was given the designation 750 Monza.

Its body was designed by Pinin Farina in Turin, but moulded by Scaglietti in Modena. The aerodynamic hood behind the driver's seat is striking. The idea for this goes back to Dino Ferrari. And Sergio Scaglietti was able to persuade Enzo Ferrari – who did not like this cockpit extension – to fit it to the vehicle. It became a characteristic feature of many open-top racing cars.

In the hands of private drivers, the 750 Monza achieved a series of successes. These included winning the Tourist Trophy and numerous victories in the USA. With the 860 Monza, Ferrari launched the ultimate expansion stage for the four-cylinder in 1956. During its short racing career, it won the 12 Hours of Sebring, among others, and thus played its part in winning the 1956 World Sports Car Championship.

With the Monza SP1 and SP2 models based on the 812, Ferrari launched the »Icona« series in 2018 and commemorated this successful chapter in its own racing history.

FERRARI 250 TR TESTA ROSSA

The Testa Rossa becomes one of Ferrari's most successful sports cars. The company wins races and championships worldwide with it. This included four Le Mans victories in just five years (1958, 1960, 1961, 1962).

A three-litre V12 engine is usually mounted under its long front bonnet, which is why it is referred to as the 250 TR. »Testa Rossa« (Italian for »red head«) describes the cylinder head covers of the engine, painted in red shrink-wrap paint, which produced around 300 hp, – a sensational litre output at the time. The backbone of the car, which weighed 800 kilos, was formed by a tubular lattice frame. The doors are more like dummies, but are required by the regulations. The front design of the first twelve-cylinder engines from 1957 is also striking, with their vertical openings for better cooling of the front drum brakes; the TR was not fitted with disc brakes until 1959.

The first Testa Rossa dates back to 1956, but was given the designation 500 TR. The two-litre four-cylinder engine from the 500 F2, which had been crowned with two Formula 1 titles, was given an increase in displacement to two and a half litres. This led to the designation 625 LM.

And the 250 TR also experienced an increase in displacement from three to four litres. The 250 TR became the 330 TR. Step by step, a new ZF steering system (1959), independent suspension and De Dion axle (1960) and fundamentally revised aerodynamics (1961) were added. Paul Frère, Le Mans winner in 1960, remembers: »The twelve-cylinder engine was simply a dream in terms of power output and smoothness.« However, its aerodynamics were so poor in 1960 that it lost out to weaker cars on the Le Mans straights. Frère's criticism bore fruit in the form of the 1961 model, with a shark's snout and high rear end à la 246 SP.

FERRARI 250 LM

The 250 LM was a big hoax, but ultimately successful. The model should actually be called the 275 LM, as with the exception of a three-litre model, only 3.3-litre vehicles were produced. However, Ferrari decided to homologate the car as a GT vehicle for the 1964 season. But as a further development of the already homologated 250 GT. Ferrari already tricked the sports commission with the 250 GTO by »selling« it as an evolution of the homologated 250 GT Berlinetta. When Ferrari tried to do this again with the 250 LM, it was too much. During his visit to the factory, the FIA commissioner only counted the 34 vehicles available. The sports authority's judgement was accordingly: A mid-engined sports car is a different vehicle to a front-engined Berlinetta. And the proven number of vehicles is not sufficient for GT homologation. Accordingly, the 250 LM is categorised as a prototype. And the 250 GTO still has to make the cut.

In fact, the racing coupé shown for the first time in Paris in 1963 was directly related to the 250 S. But Enzo Ferrari was incensed and flexed his muscles at the end of 1964. The reason for his protest was that the Italian sports authority CSAI had not supported Ferrari sufficiently in the homologation of the 250 LM. Ferrari therefore officially withdrew its racing team from motorsport. And its racing cars will compete in the last F1 races of the season in US colours under NART designation. Ferrari does not deny that the company did not produce the 100 cars required for homologation. However he accuses the officials of not having counted them quite so meticulously beforehand...

In sporting terms, however, the 250/275 LM does get its accolade. The Le Mans victory in 1965 is the most prominent in the gallery of victories stretching back to 1968.

With the Dino, Enzo Ferrari created a memorial to his son who died at a young age. The six-cylinder models open up new potential for the company - on the race track and in terms of customers

FAMILY AFFAIRS

The Dino 206 S is often referred to as the »little brother« of the 330 P3. In fact, the bodies of the two sports cars, which were assembled together in Maranello, came from the workshop of Piero Drogo. While Ferrari originally intended to put four 330 P3s and 50 Dino 206 S cars on the road for the 1966 season, the ongoing strikes that hit Italy's metalworking industry hard meant that in the end only three 12-cylinder prototypes and 18 six-cylinder vehicles were built. Ferrari even cancelled its participation in the Italian GP due to the strike.

Due to the small number of cars, the six-cylinder Dino 206 S failed to achieve homologation in Group 4, meaning that the two-litre mid-engined sports cars started in the prototype category. While the large twelve-cylinder sports car can play to its strengths on fast circuits, the compact Dino 206 S is better suited to winding courses and hill climbs. At 387 x 168 x 96 centimetres, it is short, slim and flat. And its curb weight is just 580 kilos, making light work of the 220 hp.

»There were compelling reasons to build smaller models. Ferrari developed the six-cylinder Dino 206 GT model because the regulations for Formula 2 at the time stipulated that the engine block used had to come from a car of which at least 500 units were produced. However, Ferrari only built around 700 cars a year at the time and would never have been able to accommodate the required number of engines in its own production programme, even with the new model,« explains ex-Ferrari sales manager Carlo Tazzioli. »Fiat helped by incorporating this engine into the Dino models.«

»Another reason for the expansion of the programme with smaller cars: Ferrari needed more money to finance its racing activities,« says Tazzioli. »The proceeds from series production were always used for this purpose. However, this was no longer sufficient, so production had to be increased. The Dino series was then the first in Ferrari history to exceed a production figure of 1000.«

Five decades earlier, Enzo Ferrari lost both his father and his brother, who was two years older, during the war in 1916. Both were named Alfredo, nicknamed »Dino«. And they loved to talk about cars. Enzo soon joined in. Enzo married in 1923 and his wife Laura gave him a son in 1932, who was also named Alfredo. His nickname is also »Dino«. Dino becomes his father's pride and joy and prepares to follow in his foot-

A strong design element of the dynamically sweeping body of the open-top sports car is the large rollbar behind the cockpit. It gives the car – when viewed from the side – a coupé-like appearance

Lightweight construction with style — the workplace is completely trimmed for efficiency. Wrinkle-paint dashboard and pipe-stitched seats nevertheless lend fashionable accents

Not least thanks to the compact V6, the Dino is a lightweight. And the centre of gravity is clearly in the middle of the vehicle. Good to see: the inboard brakes

Drivers can extract maximum power from the six-cylinder engine right up to 9000 rpm

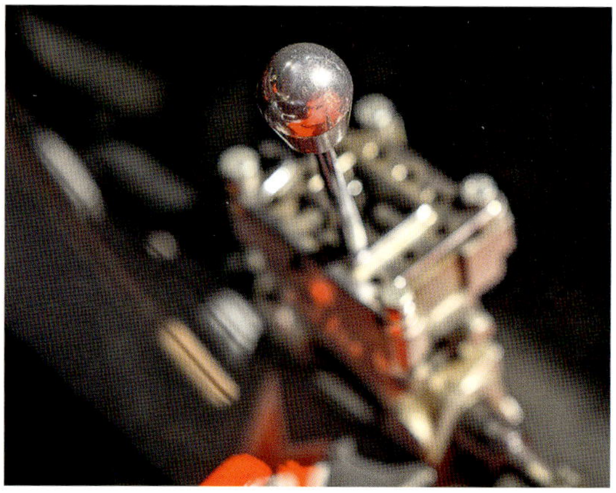

The beauty of function — a polished aluminium ball sits enthroned on the gear stick

steps. He completes a degree in engineering and spends most of his time at the company. His ideas and suggestions flow into development and production. This included the design of the V6 engine, the sound of which he would unfortunately not live to hear.

The name Dino has been used at Ferrari since 1957. However, it was originally used in connection with the six-cylinder engine, the creation of which Enzo Ferrari attributed to his son »Dino«, who died of muscular dystrophy in 1956. In fact, veteran engineer Vittorio Jano more than just assisted with the design. But for Enzo Ferrari, the new engine became a legacy and a memorial to his beloved son Alfredo.

Over the course of the years, a series of Ferrari vehicles were created, whose designation Dino refers to their six-cylinder engine. In 1965, Pininfarina presented the first prototype of a six-cylinder road sports car. It was based on the Dino 206 S and paved the way for the small mid-engined sports car 206 GT intended for road use. The production car incorporates numerous elements of this and two other concept cars. It was presented in 1967 under the brand name Dino – customers would, of course, look in vain for a Ferrari logo on the new sports car. The little brother of the Ferrari V12 models nevertheless proves to be a success. The company tapped into new target groups and expanded its production volume.

By this time, the charismatic Enzo Ferrari had long since expanded his family circle. Throughout his life, Ferrari was very fond of women. Until her death in 1978, he maintained a close bond with his wife Laura and shared responsibility for the company he had built up. But Ferrari was, in some respects, a dark horse. He had a liaison with a wartime employee, Lina Lardi, from whom his second son Piero was born in 1945. However, out of consideration for his wife Laura, Enzo Ferrari did not adopt his illegitimate son Piero Lardi until some time after her death.

From June 1966, the open-top Dino 206 S was primarily used by the Scuderia Sant'Ambroeus in hill climbs in Italy. The Milan racing team was formerly co-founded by the Zagato brothers. At the wheel is textile merchant and racing amateur Edoardo Lualdi-Gabardi, who previously drove the 196 SP, which is also on display at the Nationales Automuseum today. The result in 15 races: six overall and four class victories. From 1967, the vehicle passed through several hands. It remains successful, mostly on the mountain, but can also be seen at the 1000 km of Monza. In the meantime, the Dino has a flat spider body without the characteristic roll bar. In 1969, Lualdi-Gabardi acquired vehicle #016 once again, together with the Austrian Egon Hofer. The car changed hands thence in 1972 via a small advert in »auto, motor und sport«. The price for what was then a disused racing car: DM 8,000. In the 1990s, it then appeared

The 206 S is considered the little brother of the 330 P3. While the twelve-cylinder models bear the family name, the six-cylinder models now run under the Dino brand, the name of Laura and Enzo Ferrari's son

in the hands of a Swede in historic motorsport before being extensively restored by the factory in Maranello in 2013.

The Dino 206 S has its steering wheel on the right-hand side – just like the 125 S, 166 MM and all other racing sports cars from Maranello. The reason for this is that most racetracks are driven in a clockwise direction and therefore have more right-hand bends than left-hand bends. And the seat position on the inside of the bend offers the driver a better view and ensures a more favourable weight distribution. Porsche did not adopt this detail for its racing sports cars until the 907, which was introduced in 1967. In view of the duel with Ford, which was primarily fought with the »big brother« 330 P3, and the new three-litre Type 312 F1 racing cars, the Dino 206 S fell somewhat out of focus. Wrongly so: on the one hand, it is good for victories and titles, and on the other, it provides an excellent example of the theory that beauty is the result of form and function.

Ferrari now also uses light-alloy rims, in the remarkable 7.00 x 13-inch format at the rear

Ferrari ultimately produced only 18 of the originally planned 50 Dino 206 S. Some became concept cars

Yellow wheels with bright red bodywork – the two typical Ferrari colours

DINO 206 GT

The path to the first mid-engined Ferrari for the road is forged via three show cars presented by Pininfarina: in Paris in 1965, the coachbuilder presented the first Dino Berlinetta Speciale, followed by the Dino GT Berlinetta at the Turin Motor Show the following year and a Dino Competizione with gullwing doors and driver-adjustable spoilers at the IAA in autumn. (Formally, the 365 P Berlinetta Speciale created for Gianni Agnelli in 1966 was also the inspiration, with three seats arranged side by side and a steering wheel installed in the centre).

They provide a foretaste of the Dino 206 GT, which went into series production from 1968. The production Dino had a transversely mounted drive unit. In 1969, the displacement of the Dino was increased and it became the 246, with engine output rising from 180 to 195 hp. From 1972, an open-top GTS version was added; in keeping with the fashion and safety requirements of the time, it was designed as a »Targa« with a roll bar. This brought Ferrari closer to Porsche's top models; the 911 S with a 2.4-litre six-cylinder engine achieved 190 hp. But: in 1969, the little car from Ferrari cost DM 43,700, while a 2.2-litre 911 S, also with 180 hp, cost DM 27,140.

The Dino also made its television debut, including roles in »The Persuaders« with Tony Curtis and with Robert Wagner in »Hart to Hart«.

Even though the Dino comes from the newly expanded factory in Maranello in 1970, it is not known as a Ferrari as it is marketed under the brand name Dino. The two-seater 246 was followed in 1974 by the 2+2-seater Dino 308 GT4 and the 208 GT4, which was intended for the Italian market with a reduced-displacement two-litre V8. Beneath its bodywork, exceptionally tailored by Bertone, was technology that would cause a sensation in the 308 presented in 1975. This would also bear the name Ferrari.

LANCIA STRATOS

From 1967, the rule for Formula 2 was that the 1.6-litre engines used there had to be taken from a vehicle of which at least 500 units had been produced within a year. Enzo Ferrari always kept an eye on this category, but Ferrari was unable to fulfil the homologation requirements with its exclusive production numbers. This led to an agreement with Fiat in the spring of 1965: Ferrari would supply Fiat with engines for sporty Fiat models from the end of 1966, which would then be produced in sufficient numbers. Specifically, this involved the Fiat Dino Spider (1966-72, 1546 units) designed by Pininfarina and the Fiat Dino Coupé (1967-72, 6037 units) with Bertone bodywork.

The engine was based on the Ferrari Dino racing engine designed by Alfredino (with the help of Vittoria Jano). Former Ferrari engineer Aurelio Lampredi came into play in the development of a road-going offshoot. He simplified the valve and chain drive and switched from aluminium to grey cast iron for the engine block.

From 1972, almost 500 examples of the Lancia Stratos, which was uncompromisingly designed for rallying (with space for helmets in the doors), were produced. This rewarded the Lancia brand, which had also belonged to Fiat since 1969, with three consecutive World Rally Championships (1974-76).

While the Dino V6 is mounted longitudinally in the front of the two Fiat models, in the Stratos it lives transversely in a mid-engine design under the wedge-shaped body. The 270 hp of the competition version meets a vehicle weight of less than 1000 kilos! As part of the Dino project, Fiat sent senior managers from Turin to Maranello. One of them was Pier Ugo Gobbato – son of the former Alfa boss (1888-1945). He finally negotiated the deal in which Enzo Ferrari handed over half of his company to Fiat in 1969.

FERRARI 308 GTB

The curtain rises on the 308 in Paris in 1975, allowing Ferrari to occupy the market segment below the twelve-cylinder super sports cars and open up the brand to new clientele. The lines, for which Pininfarina head designer Leonardo Fioravanti was responsible at the time, still epitomise the spirit of the 1970s today. At the time, it proved to be so timeless that it remained relevant well into the 1980s.

»It was not foreseeable that the success of this car, especially in the USA, would be so gigantic,« recalls Ferrari salesman Carlo Tazzioli. Accordingly, Ferrari fundamentally reorganised its production once again: By 1977, Scaglietti had produced the first 308 with a fibreglass body (Vetroresina). This proved to be unsuitable for series production. Ferrari subsequently invested in moulds so that the 308 and 328 models could be manufactured from sheet metal.

During its 15-year production period, the eight-cylinder mid-engined model underwent a series of updates. In 1977, the GTB coupé was joined by the GTS »Targa« version; in 1980 fuel injection replaced the carburettors; followed by the introduction of four-valve technology (quattrovalvole) in 1982; and in 1985 – with the 328 model – an increase in displacement to 3.2 litres, resulting in 270 hp.

The 308 GTS, which TV private detective Tom Selleck alias Thomas Magnum drove on the streets of Hawaii in the hit series »Magnum, p.i.« from 1980 to the present day, is undoubtedly the world's best-known model. Since 2022, it has even been available as a Playmobil toy model – Mini-Magnum.

The 330 P3 is one of the absolute icons in Ferrari's illustrious history. In addition to its beguiling beauty, it also stands out for its role in the fierce competition with Ford

THE SPEEDY HOLY GRAIL

Although the 250 GTO is considered by many to be the ultimate, the much rarer 330 P3/412 P and 330 P4 have the role of the Holy Grail among Ferrari racing sports cars. The bodywork realised by Piero Drogo is a feast for the eyes and has been copied or reinterpreted many times to this day. This ranges from the La Ferrari to the Ferrari P4/5, which Pininfarina built in 2006 based on the Enzo for US billionaire James Glickenhaus, to the Ferrari Daytona SP3 from 2021.

The history of the original from 1966 and 1967 is already exciting, as the 330 P3/4 was created at a challenging time: Ferrari produced just over 650 cars a year in the mid-1960s. At the same time, the gap between road and racing cars was widening. This meant that Enzo Ferrari's calculations no longer worked out: Selling sports cars to the (very) well-heeled few was hardly enough to finance the company's extensive sporting activities. After the early death of his son Alfredo, Enzo Ferrari – now well over 60 years old – also endeavoured to secure the future of the company he had founded.

In 1963, an offer from Ford seemed enticing. The US car giant wanted to adorn its mass-produced cars with the glamour of the Italian thoroughbred. After initial negotiations, the idea was finalised: road models as Ford-Ferrari, racing cars as Ferrari-Ford. It could have happened that way but for the fact that Enzo Ferrari was bothered by a subsequently amended clause in the contract that would have curtailed his motorsport decision-making powers. The tablecloth was figuratively speaking cut, yet the contract was cancelled. Ferrari would later join forces with Fiat. But before that, Henry Ford II was out for revenge: if they didn't win together, Ford would beat Ferrari. This was the starting signal for the Ford GT40 and the years that will retrospectively go down in history as the »Ford-Ferrari war«.

The 330 P3 is Maranello's answer to the GT40. Its lightweight aluminium body conceals a filigree tubular frame and a further developed four-litre V12. Thanks to petrol injection and dual ignition, the engine now produces 420 hp. In addition, there is a 5-speed gearbox from

The beautiful aluminium body of the racing car was created in Piero Drogo's workshop and blends harmoniously into the Ferrari design language. The 330 P3 is significantly flatter and wider than its predecessors

Ferrari #0844 was born in 1966 as the 330 P3 and advanced to the 412 P the following year. Over the course of its career, it changed its appearance several times. Today it presents itself as it did in Daytona 1967

The steering wheel rim is leather, the spokes are non-reflective black. And on the short gear lever sits a non-slip knob, which the driver uses to shift through the five gears

ZF, which will later be replaced by an in-house design. For the chassis, the team utilises the in-house Formula 1 department. The new frame reduces the vehicle weight and has a positive effect on the rigidity of the racing car. All powerful arguments in terms of competitiveness. At 720 kilos, the 330 P3 is also significantly lighter (100 kg) than its predecessor, the 330 P2. The proportions of the 95 centimetre flat and 1.78 m wide racing sports car are also impressive.

However, strikes in the Italian steel industry hampered preparations for the 1966 season. Instead of the usual five vehicle chassis, only three were produced this time (#0844, #0846 and #0848). The Loh Collection vehicle bears the number #0844 and contested a total of eleven races in 1966 and 1967. The drivers at the wheel were all racing greats: John Surtees, Mike Parkes, Nino Vaccarella, Lorenzo Bandini, Ludovico Scarfiotti, Jean Guichet, Pedro Rodriguez and Giancarlo Baghetti.

In its debut year, the (naturally) red-painted works car set highlights with victories in the 1000-km races in Monza and Spa-Francorchamps. At the Targa Florio, Vaccarella/ Bandini were in the lead when an unsuccessful overtaking manoeuvre ended their race in the ditch on lap 7. At the Nürburgring, the 330 P3 in the hands of Surtees/Parkes/ Bandini set the tone with the fastest time in practice. Nevertheless, Ferrari race director Eugenio Dragoni favours the more compact Dino 206 S for the race. The highlight of the season is undoubtedly Le Mans. But not

In this guise, #0844 conquered the podium of the Daytona 24h on 5 February 1967. In June, it wears a white and blue livery at Le Mans. And from September, it raced in the CanAm series – as a spider with a flat bonnet and no roof

The beauty of uncluttered engine architecture – the harmony of powerful performance and Italian style

Operating the six carburettors requires a strong right foot

for Ferrari: the last 330 P3 (#0844) remaining in the race retired before dawn – engine failure.

For the 1967 season, the 330 P3s received a technical update and were handed over to privateer teams. Weber carburettors are fitted again instead of the mechanical Lucas injection system. This is accompanied by their new relationship: 412 P. Meanwhile, the factory opts for three brand-new 330 P4s. They differ from their predecessor with a modified chassis and three valves per combustion chamber. This ended the factory career of #0844; the car's sporting career, however, continues. It goes into the

The light blue wheels are a recognisable feature of the NART team. At the start in Le Mans, #0844 shone in white and blue. From 1968, the national colour scheme was no longer compulsory

The fresh air vent provides a little comfort. The toggle switches are also used by Alfa Romeo

The silky matt black look gives a foretaste of the automotive fashion of the coming 1970s. At the time, only a few engines ventured into these rpm regions

For aerodynamic reasons, the large bonnet has a long, flat rear window. Nevertheless, the engine is hidden from view under the thin aluminium skin

hands of the California-based North American Racing Team (NART for short, officially N.A.R.T). The team owner is Luigi Chinetti – the former racing professional and Le Mans winner is a Ferrari importer in France and the USA.

At the start of the season, #0844 (Rodriguez/Guichet) secures Ferrari a magnificent one-two-three finish at the 24 Hours of Daytona with a third place. The victory at the home ground of the US competition means a lot to Ferrari. Its significance is also reflected in the model designation for the Berlinetta 365 GTB/4 »Daytona« presented in Paris in 1968. After another start in Monza (retirement following an accident), the car was given a white and blue livery for its Le Mans outing (in accordance with the ACO regulations) – the national racing colours were still mandatory for a US team that year. While the 330 P4s named by the factory finish in second and third place, the Le Mans appearance for the NART team's starting number 25 ends in the night, once again with engine problems.

Again, the car changed its appearance: in Maranello, #0844 was modified into a Group 7 Spider – partly due to the three-litre regulations that came into force in 1968. With its open body and 4.2-litre engine, the car was well equipped for the North American CanAm racing series. At the end of 1967, it contested two more races there, before rolling into retirement. As an »all-time great« in Ferrari history, it travelled through well-known collectors' hands in the USA, Japan and Great Britain over the following decades to become one of the highlights of the Loh Collection in 2011.

FORD GT40

Without Ferrari, the Ford GT40 would hardly have been created. Maranello was navigating difficult economic waters in the early 1960s. A strong partner would have been just what was needed. For the US mass manufacturer, acquiring the prestigious brand as an image-building measure would be a very attractive option. Via the Ford subsidiary in Cologne, the Detroit management received a tip that Ferrari was on the lookout for a new partner. However, after Ferrari had bought the building of the Italian Ford subsidiary in Bologna years earlier, the companies had already made initial contact.

Ultimately, the parties did not come together and Ford responded with the most expensive and extensive sports programme in motorsport history to date. The company knows no bounds when it comes to developing the mid-engined racer. In addition to former Aston Martin team boss, John Wyer, Ford also secures the services of Carroll Shelby. Computers are also used for the first time. In June 1964, the first car celebrates its premiere at Le Mans. However, it took another two years of development before the first Le Mans victory in 1966. In 1965, Ford celebrated the first victory of a GT40 (Gran Turismo, 40 inches high) at Daytona. Over the course of its career, Ford went on to win at Sebring, Montlhéry, Silverstone, Spa, Reims, Hockenheim and many other circuits. By far the most valuable, however, are the triumphs that the GT40 achieved at Le Mans between 1966 and 1969. Ford thus succeeded in making the unimaginable conceivable: The dominance experienced by Ferrari until 1965 is broken. And Ferrari finally made a lifelong alliance with Fiat.

PORSCHE 906 »CARRERA 6«

The Carrera 6 quickly becomes a boyhood dream. No doubt in part because the sports car is takes centre stage on the »Carrera« tracks set up in numerous children's rooms. The counterpart to the white Carrera, in the gift packs from the manufacturer of electric slot tracks of the same name, is offered in the red colour of the racing cars from Maranello. But model car manufacturers such as Dinky Toys and Märklin also downsized the mid-engined speedster, which succeeded the 904 Carrera GTS from 1966.

The 906 or – remember the name dispute with Peugeot – Carrera 6, is uncompromisingly tailored to racing. Behind the cockpit is the further developed six-cylinder boxer engine from the then also new 911.

As for the chassis, a tubular frame replaces the box frame construction of the Carrera GTS – it is stiffer and lighter. Gullwing doors, as first seen on the Mercedes-Benz 300 SL, caused a sensation. An aerodynamically favourable long-tail version is also available for use at Le Mans.

The factory only entered the 906 in a few races, as the team led by Porsche family member Ferdinand K. Piëch was constantly developing new sports car prototypes in quick succession. After successes in Sebring, at the Targa Florio, on the Nürburgring and at Le Mans, the 906 won the World Sports Car Championship in Division II (up to 2 litres displacement), and in the hands of privateers, the 60 or so vehicles produced collected no fewer than 287 overall victories by 1978. The number of class victories achieved in 2226 races is almost impossible to calculate.

FERRARI 275 GTB/4

With the 275 GTB, Ferrari presented the first V12 Berlinetta with independent suspension and transaxle construction at the 1964 Paris Motor Show. 24 months later, the model was updated both visually and technically. In Paris, in 1965, a revised body with a larger rear window was on show. In Brussels in 1966 the more harmonious version, known as the »long nose«, was presented. Again in Paris, in that same year, it's aired – now with the revised 60° V12, with a total of four camshafts. The testers from »Car & Driver« wrote at the time: »While the noise of the valve train in the old single-camshaft engine had almost drowned out the exhaust noise, the mechanically very quiet dohc unit allowed the driver to completely surrender to the inimitable exhaust note.«

The appearance of the two-seater, designed by Pininfarina and built by Scaglietti, is reminiscent of the 250 GTO thanks to the covered headlights, the side gills and the spoiler lip at the rear. With an official 300 hp, the (DM 52,900) GT is just as powerful as the successful racing model. The »Car & Driver« testers describe their driving experience as follows: »Everything about the GTB is designed for a brisk pace. The light flywheel and the hard-gripping clutch do not forgive any mistakes when changing gear – the car bucks even if you make a mistake of just 100 rpm. This Ferrari requires the kind of extreme concentration that is only possible at high speeds.«

Exactly 442 vehicles of the 275 GTB were built between 1964 and 1966, and a further 330 of the 275 GTB/4 produced in 1966 and 1967. Ferrari also supplied the Berlinetta with an aluminium body for sports drivers on request.

THE KING AMONG KINGS

In the Schumacher era, the Scuderia achieved ground breaking successes, which, to this day, haven't been replicated. Between 1996 and 2006, the team collected six constructors' titles, five drivers' championships, 87 victories and 72 pole positions. The successes of 2007 and 2008 are still based on the previous work of the »Dream Team«

When you think of Formula 1, Ferrari inevitably comes to mind. Enzo Ferrari knew Grand Prix racing and its surroundings from his time with Alfa Romeo in the 1920s. In the 1930s, his team was already one of the key protagonists in the highest motorsport category. And so the leap into Formula 1, which was launched in 1950, seemed a natural one for the young automobile brand. The red racers literally became at one with Formula 1 – in successful and less successful years.

In seven and a half decades, the Scuderia collects 16 constructors' titles, and with Ascari, Fangio, Hawthorn, Hill, Surtees, Lauda, Scheckter, Schumacher and Räikkönen, nine drivers manage to win a total of 15 drivers' titles. Michael Schumacher achieved 72 victories in 181 races for Scuderia, while Niki Lauda brought his »rival« 15 successes. Sebastian Vettel achieved 14 and Alberto Ascari 13 F1 successes in red. Felipe Massa and Fernando Alonso have 11 each. In total, the list of drivers at the wheel of a Formula 1 Ferrari who have climbed to the top step of the podium at the end of a race comprises 40 names. The list of those who have unsuccessfully tried is much longer.

For decades, Ferrari has benefited from the fact that its racing team exerts a unique fascination. Enzo Ferrari can pick and choose his drivers – and, as a consequence, pay them less than competing teams.

Formula 1 experiences highs and lows. The downsides include, first and foremost: tragic deaths among the racers, track marshals and spectators alike. The motorsport premier class has had to overcome difficult economic times as well as disputes between the world governing body, the constructors' association and the organisers. Formula 1 garners strength when the interim Brabham team boss, Bernard Charles Ecclestone, bundles up the interests of the teams and negotiates for entry fees and television rights on behalf of the FOCA (Formula One Constructors' Association). This also gives the other teams the weight that Enzo Ferrari has always proclaimed for his racing team. »The Italian was proud of the fact that negotiations in racing were in his blood. Even before Ecclestone was born, he had transported his cars to the Italian border and refused to enter France until he had received a decent entry fee from the organisers of the race in Monaco – in cash,« describes Ecclestone biographer Tom Bower.

Television and the associated broadcasting rights brought Formula 1 into living rooms around the world. Within a short space of time, it became an international event on the scale of the Olympic Games or the Football World Cup – except that the races took place every 14 days. To this day, Formula 1 continues to reinvent itself, while remaining true to itself. It uses social media and new communication channels, such as the Netflix documentary »Drive to survive«, to tap into new target groups. As a result, it remains attractive to young people and has long since conquered the US market. In its early days, Americans and the European-influenced F1 racing circus were still at odds with each other.

Formula 1 enters its 75th season in 2024. And Ferrari is the only team that can look back on an unbroken history in the premier class.

Eddie Irvine won the opening round of the 50th Formula 1 season in Australia in 1999

In the 1980s, pit stops became part of the F1 dramaturgy

FORMULA 2

When the young motor journalist Paul Frère visited Ferrari for the first time in 1948, Enzo personally showed him »how the 1.5-litre supercharged engine was to be installed in the same chassis for Formula 1 races and the 2-litre version without a supercharger for future Formula 2 races.« Ferrari's first model, the 166 F2, actually won 13 of a total of 17 races in 1949. Enzo Ferrari always had his eye on the junior series. Formula 1 resorted to Formula 2 designs twice: in 1952 and 1953 due to a lack of attractive cars in accordance with the major regulations, and in 1961 to slow down the now breathtakingly fast monoposti. Both times, the Scuderia knew how to take advantage of the situation. First with the 500 F2, later with the 156.

Like Formula 1, Formula 2, the lower-level category, experienced numerous changes to the regulations, good years and not so good years. The 1.5-litre V6 developed for the F2 in 1956 plays a major role in Ferrari's history, as Enzo Ferrari attributes its development to his son Alfredo »Dino«, who died in 1956. As a matter of fact, Ferrari veteran, Vittorio Jano, lent the young engineer a helping hand.

However, the Dino engine showed great potential and could be further developed and even industrialised for series production well into the 1960s. Fuelled by the V6 design, ambitious young drivers competed with renowned professionals from the top leagues of motorsport in the 1960s and 70s – enjoying competitive racing and attractive winner's purses.

INDYCAR / CART / INDIANAPOLIS

Enzo Ferrari took an early look at what was happening on the other side of the pond. In 1936, Tazio Nuvolari won the Vanderbilt Cup held at the Long Island Airport Circuit in an Alfa Romeo 12C-36 monoposto, entered by Scuderia Ferrari. His team-mate, Antonio Brivio, sets the fastest time in practice and finishes third.

From 1950 to 1960, the 500 Miles of Indianapolis is recognised as the US round of the F1 World Championship. However, participation was not compulsory for the teams active in Formula 1, and so the US teams remained among themselves in this race, which was already legendary at the time. Only in 1952 did Ferrari send Alberto Ascari and a 375 prepared for the oval race to Indianapolis. The red racing car, adorned with the »Ferrari Special« inscription, actually led the race for a long time, but then lost a wheel due to hub damage and retired after 40 (of a total of 200) laps.

In the mid-1950s, the Scuderia sent its F1 monopostos to the races in Argentina, but they were no longer to be seen at the Indy 500.

In the second half of the 1980s, the topic of IndyCar in Maranello picks up speed once again: At the same time as FISA was considering adapting the F1 engine regulations and banning the use of twelve-cylinder engines, Enzo Ferrari had a racing car developed for the CART IndyCar series, which was important in the USA. Under the direction of designer Gustav Brunner, the »Ferrari Formula Cart« equipped with a V8 turbo (internal project designation: Ferrari 637) was created in 1986. However, when FISA abandoned the idea of using turbo four-cylinder engines in Formula 1 and instead allowed 3.5-litre naturally aspirated engines to be used as V12s, the ready-to-run US racer disappeared into Ferrari's history as a footnote.

HILL CLIMB

Competition on the mountain knows no wheel-to-wheel duels. Driver and car race against the clock, gravity, the track and the mountain. Once on the move, the driver is left to his own devices, his knowledge of the route and his ability to find the ideal line through woods and fields, always narrowly avoiding the abyss – hill climbs are like a balancing act on a tightrope. No other racing discipline is more demanding in terms of driving skills. At the same time, amateurs can also compete with the professionals here, categorised in different classes. This makes it (relatively) inexpensive to get started. In the professional camp, manufacturers such as Porsche and Ferrari spare no effort. New lightweight prototypes with components made from exotic materials are constantly being created.

Until the 1970s, there were numerous hill climb races in many regions. They were easier for the organisers to stage than circuit races. From the second half of the 1950s onwards, hill climb sport experienced its heyday. For this reason, a minimum course length of five kilometres applies to events that want to count towards the European Hill Climb Championship – if the course is less than ten kilometres long, it must be completed twice. In addition, the difference in altitude between the start and finish must be at least 350 metres and the average gradient at least 5 percent. Gaisberg, Schauinsland, Rossfeld, Trento-Bondone, Fléron, Mont Ventoux and Ollon-Villars are some of the European Championship venues. It is only with increasing organisational and safety requirements that the events fall through the cracks, one after the other.

Formula 2 is a popular playground for budding F1 drivers. And even for the Grand Prix aces, starts in this category are interesting because they are lucrative. The same applies to the Tasman series held »Down Under«

GRAND FINALE FOR »LA PICCOLA«

After the Formula 2 models 166 F2 (1948), 500 F2 (1952) and 553 F2 (1953), the first Dino followed in 1957. It was called the 156 F2 and featured the 1.5-litre V6 developed by Vittorio Jano and Alfredo »Dino« Ferrari. The engine was christened Dino in memory of the son of Enzo and Laura Ferrari, who died in 1956 at the age of 24. The 65° V6 became world-famous due to the fact that it was used in a number of other sports and racing cars. And it even became a central element of the brand, enabling Ferrari to successfully launch new models below its twelve-cylinder sports cars. In addition, there are high-volume Fiat models that also bear the name Dino. And this, in turn, also had something to do with Formula 2...

While Ferrari was still able to fulfil the homologation requirements for Formula 2 in the 1950s, it was no longer possible for the small series manufacturer to meet the number of units required from 1967 onwards: FISA requires at least 500 units to be produced for F2 use. This was impossible for a manufacturer that produced a total of 619 cars in 1965. But it was important to Enzo Ferrari to »maintain technical contact«. And so the »Ingegnere« managed to reach an agreement with Fiat at the beginning of 1965. The Turin-based company took over the production of the engine developed by Ferrari and ensured a decent production volume of road vehicles with a Spider and a Coupé. This support from the future parent company cleared the hurdle for the use of the Dino 166 F2, which appeared in 1967.

The »Piccola« with chassis #008 is enjoying a second career »Down Under« after its stints in the hotly contested Formula 2 European Championship. And what a career: It has won two titles in a row

At the time, Formula 2 races were very popular with drivers and teams as well as spectators. It was not only the fans who liked the fact that well-known F1 professionals competed with young drivers in the dense starting fields. Numerous Formula 1 careers are built on the solid foundation of F2 successes. For motorsport photographer Ferdi Kräling, Formula 2 is just as much a part of the calendar as Formula 1, the World Sportscar Championship and the European Hillclimb Championship: »The Formula 2 races were exceptionally good, with huge starting fields. Nice people, great times. And that's what matters in the end.«

However, a large part of the 1967 season elapsed before Ferrari entered the fray again; there was only one race (retirement for Jonathan Williams in Rouen). Improvements were made in Maranello: instead of three valves, the 1.6-litre V6 was fitted with four-valve heads for the following season. Basically, this was a further development of the 1961 F1 engine, which in

The compact shift gate requires a precise effort from the driver on the sculpted shift lever

turn was directly descended from the original Dino engine. In 1968, the engine of the »piccola Ferrari« now produced 210 hp at 10,500 rpm.

New Zealander Chris Amon achieved fifth place in his first race at Hockenheim. In Enna-Pergusa, Ferrari drivers Brambilla, Bell and Ickx finished third, fifth and sixth respectively. Ultimately, it took until October for Ferrari to add its name to the winners' lists of the hotly contested F2 European Championship: Vittorio Brambilla wins in both Hockenheim and Vallelunga. Derek Bell (third in Hockenheim) and Andrea de Adamich (second in Vallelunga) rounded off the successes of the Ferrari Dino F2 166.

The 1966 Le Mans winner (in a Ford GT40) joined Ferrari in 1967. »As Amon had no idea of the Scuderia's internal affairs and miserable salary structure, he was content to drive without a fixed income and for 50 per cent of the prize money,« says Ferrari biographer Brock Yates. In a sports car, this paid off for the son of a wealthy sheep farmer: the New Zealander won at Daytona in 1967 (at the wheel of a 330 P4). In the Formula 1 Ferrari, on the other hand, he seemed to be downright unlucky: He was often in the lead, but then repeatedly retired due to technical faults. Amon achieved five pole positions and 850 kilometres in the lead. But to this day, he is considered »the best F1 driver who never won a race«.

In 1969, Ferrari equipped the Formula 2 racer with a 2.4-litre engine and sent the monoposto, which had been further developed into the 246 T, to the other side of the world. At the time, the Tasman Cup, held in Australia and New Zealand, was very popular with racing professionals. Here, lavish starting and prize money can be won. In addition, the events held from the beginning of January

The design of the monoposto is reminiscent of a dugout canoe. The aerodynamic finishing touches in the form of the front wing and the attached rear wing are still unfinished

to mid-February are good testing opportunities, free from conflicts with the usual racing calendar. Ferrari is sending Chris Amon and Derek Bell »Down Under«. With four victories (New Zealand GP Pukekohe, Levin, Australian GP Lakeside and Sandown Park) and two third places, Amon confidently secured the title at his home race. In doing so, he puts Jochen Rindt and Piers Courage in their place. The following year, New Zealander Graeme Lawrence drove the now privately entered Dino 246 T with chassis number #008, winning the season opener in Levin, New Zealand, and subsequently giving Ferrari its second title in the series.

After competing in Formula 2, the »piccola Ferrari« modified into the 246 T wins the Tasman series with aplomb

The aerodynamically shaped rear-view mirrors are each enthroned on three ornate struts. They are screwed on to the aluminium skin

In addition to the spring/damper unit, parts of the rear suspension and a drive shaft can also be seen here

Fine-meshed fly screens prevent foreign bodies from entering the six combustion chambers via the stacks

Engineering of the pre-computer era: the rear wing mount looks just as masculine as the visible rivets in the body panels – evidence of uncompromising lightweight construction

A total of seven 166s are built at Maranello, three of which are converted into the 246 T. The chassis of the 166 and 246 T consisted of an aluminium-clad tubular frame, as was customary for Ferrari at the time. The exquisite car weighed just 425 kilograms with the smaller engine, compared to 440 in the Tasman series. The 166 reached speeds of up to 265 km/h with 210 hp. In the 246 T, 285 hp is then available, enabling speeds of up to 300 km/h.

Incidentally, Ferrari, like other F1 racing teams, is now focussing on promoting young talent in the lower formula classes. However, the Ferrari Driver Academy, founded in 2009, concentrates on supporting and promoting the drivers; Maranello is not technically involved. Mick Schumacher and Charles Leclerc are two of its graduates. Both also have a Formula 2 title under their belt (2020 and 2017).

The inclination of the front wings attached to the side of the radiator opening can be varied. Hot air escapes from the radiator through the openings above the start number

FERRARI 512 S

While Porsche sent its five-litre »production« 917 racing sports car into its first race at Spa-Francorchamps in Belgium in May 1969, Ferrari announced at the end of April that it would soon be playing in this league too. The 512 S, a five-litre, twelve-cylinder car, also appeared in 1970. Enzo Ferrari invested some of the money he had earned from the sale of company shares to Fiat in this model.

In January 1970, Ferrari recreates the previous year's photo motif shot in the Stuttgart factory. The red racing cars are lined up in a row. However, only 17 complete cars are built, with Ferrari providing the necessary individual components for a further seven. That makes a total of 24 instead of the required 25 – but the sports authorities still give the go-ahead.

In Le Mans, no fewer than eleven 512 S cars enter the race, plus one 512 P. Half a dozen are works cars. The others are entered by private teams and private drivers. Their purchase price: 24 million lire (equivalent to DM 160,000). Not all the cars find a buyer: some chassis go to Pininfarina for futuristic concept cars.

The Le Mans results for 1970 remained sobering: in the end, the cars finished fourth and fifth. The first 512 S fails in the first hour with bearing damage. An accident in hour four then eliminated four more 512 S cars in one fell swoop – and all this only after the Swedish Scuderia Filipinetti driver Reine Wisell braked unexpectedly due to the blinding evening sun. He takes his brand colleagues Parkes/Müller, Bell/Peterson and Merzario/Regazzoni out of the race. After seven hours, the 512 S of Kelleners/Loos has an accident, and in the early hours of the morning Juncadella/Fernandez and Ickx/Schetty have to retire. The retirement of Manfredini/Moretti completed the series of bad luck.

FERRARI 312 PB

The »monster« 512 S proves to be a horse that can hardly be tamed, and so Ferrari opts for a new steed for the World Sports Car Championship. The 312 PB followed a different concept: compact, flat, intricate and with the 180° V12 borrowed from the F1 racing car 312 B with 450 hp; suitable for long distances. The small front wheels measure 8.6/20 x 13 inches, which reduces the rotating mass. The rear wheels are 13.5/24 x 15.

Further development work was needed in 1971. But in 1972, the 312 PB won all the races it entered. The list ranges from the 1000-km races in Buenos Aires, Brands Hatch, Monza, Nürburgring and Zeltweg to the 6-hour races at Daytona and Watkins Glen and the 12 Hours of Sebring; including world championship titles for driver (Jacky Ickx) and constructor. The sports car successes bridged the lean period that Ferrari, its drivers and fans had to endure in Formula 1 at the time.

In Le Mans, only the 4-hour race is on the programme – at the time considered the preparation event for the actual endurance classic at the Sarthe. Ferrari was concerned that the Formula 1 engine would not be able to cope with the rigours of day, night and day. In 1973, the still successful 312 PB also took part in the 24-hour race. For a long while, it looked as if the Ferrari duo Ickx/Redman would win, before Ickx headed for the pits with engine damage 30 minutes before the end of the race. The sister car of Pace/Merzario saved second place.

After the season, Ferrari withdrew from endurance racing at the factory to concentrate on Formula 1 – until the LM and WEC comeback in 2023.

FERRARI 499P HYBRID

50 years after withdrawing from the World Sports Car Championship, Ferrari is back in the top league of endurance racing with a works entry. While the FIA's entry criteria in the mid-1970s were unfavourable for Ferrari, the regulations for the LMDh and hypercar category (LMH) that came into force in the mid-2020s opened up ideal opportunities for efficient hybrid technologies, which are also becoming increasingly important in production car manufacturing. The LMDh category is intended for use in the WEC, Le Mans and the US IMSA series. A number of standardised components – such as chassis and hybrid technology – reduce costs. A sophisticated marketing concept also enables Ferrari to structure the project as a profit centre. Ferrari not only sold the six 499P cars built for racing to loyal customers in advance, but also launched a small series of 30 vehicles six months after the Le Mans triumph. Unit price: a good 6 million euros. From then on, these vehicles were allowed to be driven at Ferrari racing events around the world as part of the Corse Clienti programme. Name of this model: 499P Hybrid Modificata. As the small-series variant is free of regulatory restrictions, the Modificata has an even higher system output (3.0-litre V6 turbo and electric motor) of up to 870 hp (compared to the 707 hp of the racing car).

Modern Ferraris offer a sophisticated electronics architecture. Ferrari acquires the expertise for this in Formula 1

TECHNOLOGY AND INNOVATIONS

Without passion, self-confidence and courage, there can be no innovation. A sustainably successful, prosperous company needs the right balance between daring to try something new and trusting in the tried and tested. »The Caesar of Speed«, as Enzo Ferrari was also known, knew how to constantly break new ground throughout his life, invest the proceeds immediately and yet not put the company in unnecessary danger.

With its firm foothold in the world's most important races, the performance of its cars and the prominence of its clientele, Ferrari was under scrutiny from the very beginning. In addition to the highlights, envious people also look for the casting shadows, which certainly exist in connection with the designs from Emilia-Romagna. »The Ferrari was always a sports car with character,« writes motoring chronicler Hans-Karl Lange. »Pronounced strengths were offset by no less pronounced weaknesses, especially in the early years. Steering as stiff as in a lorry, fragile clutch, undersized rear axle, erratic brakes, dubious electrics, stubborn handling, average road holding at best. But the engine!«

However, a look back over the brand's eight decades of tradition clearly reveals many highlights. Naturally, Ferrari benefits from the fact that the company and its engineers are constantly competing with the best of the best, once a fortnight, the world over. There is no other car manufacturer under such pressure to succeed. Enzo Ferrari knew very early on how to surround himself with the best engineers. He was proud of the honorary title of »Ingegnere«, which the University of Bologna (rightly) awarded him in 1960. He is credited with bringing the right people together in the right place at the right time. This is his greatest gift – in addition to his ability to keep a firm grip on the reins at all times and not lose sight of the goals he was striving for. Ferrari is the »story of people, machines, passion and constant striving forward,« says Luca Montezemolo. »Ferrari is a company that, throughout its history, has always looked to the future with determination and put all its love into it, and for which racing and technological progress have always been paramount. Always looking to the future and keeping a close eye on the world – these are the lessons I learnt from Enzo Ferrari,« says his former assistant, who later became Ferrari boss and Fiat CEO. Among other things, Montezemolo is to thank for a quality offensive that has had a lasting effect to this day.

When looking at the long list of innovations, the focus quickly shifts to aerodynamics as well as the engine. In true Ferrari style, the company always knows how to utilise the innovative strength of its partners and suppliers. Since the later 1980s, electronics expertise has also been added. In 1989, for example, the electro-hydraulic gearstick developed together with Magneti Marelli celebrated its premiere in Formula 1 – today there is no longer a racing car whose gears are not changed by pulling a lever on the steering wheel. Active suspensions with adjustable dampers or even electronically adjustable differentials (such as the E-Diff introduced in 2004 with the F430) also made their way into series production via Formula 1. Record-breaking driver Michael Schumacher takes great pride in transferring F1 expertise to road sports cars through tireless testing and development work. The F1-Trac-Control in the 599 GTB, which can be controlled via the »Manettino« on the steering wheel, is a good example of this. With the appearance of the Ferrari 458 Speciale, both elements were combined to form Side Slip Control (SSC). With the 488 Pista, there is the so-called Ferrari Dynamic Enhancer, in which the brakes also play their part in improving the driving dynamics.

There is no end in sight, as everything remains the same: The better is the enemy of the good. This has been Ferrari's philosophy from the very beginning.

Super shark: The Ferrari 156 from 1961 is nicknamed »Supersqualo« – because of its shark-like front end. In Formula 1 it consumes sizeable prey...

With Niki Lauda and Luca di Montezemolo, Ferrari is beginning to make efficient use of the opportunities on site – such as the Fiorano test track

Modern times: All vehicle information can be found in the cockpit display; from traction control to air and oil pressure through to navigation

In Formula 1, Ferrari has been yearning for the world championship crown since 1964. Jacky Ickx brought the Scuderia a few victories with the 312 B3, but it wasn't until Niki Lauda and the 312 B3-74 that Ferrari managed to set the course for further F1 titles

THE SIGNS POINT TO SUCCESS

1972 and 1973 are turbulent years. The energy crisis left its mark in Italy, culminating in the end of production of the Fiat Dino Coupé and Spider models. Ferrari also suffers economically – despite the windfall from its partner Fiat, which came on board in 1969. In sporting terms, the Scuderia dominated the 1972 World Sports Car Championship. In 1973, however, a series of victories and ultimately the title went to Matra. And in Formula 1, the traditional racing team became entangled in intrigues and problems of its own making.

In terms of racing cars, Ferrari favours the twelve-cylinder boxer, which is technically a 180° V12. In Monaco 1971, Ferrari launched the 312 B2 model. Although it initially brought the Reds important successes, it proved to be not yet fully developed. And when the development of the innovative 312 B3, planned for 1973, stalled, the 312 B2 remained the sports car used by Maranello until well into the 1973 season. Despite the two-and-a-half racing seasons that this car engaged, it lost out to the competition. It was not until the 312 B3-74 was introduced at the start of the 1974 season that Ferrari returned to F1 victory. After finishing second in the first two races, Niki Lauda wins race number four of the season. In addition to its design, the new team manager Luca Cordero di Montezemolo and the newly signed driver also make a significant contribution to the renaissance of the traditional racing team.

Jacky Ickx is successful for Ferrari in the 1972 and 1973 constructors' world championships. In Formula 1, he achieved a victory in the 1972 German GP. Here he is driving the museum car in Brazil in 1973

»If I could wish for a period in the history of Formula 1 in which I would like to work as a designer,« enthuses Adrian Newey, »then this would be it.« The successful designer (for March, Williams, McLaren and Red Bull) is referring to the years between 1968 and 1977: »If you look at the cars that were on the grid in the 1970s, you can see how different they all are. The regulations were not yet very extensive, and as a designer you could experiment to your heart's content. However, colleagues back then had little idea what effect this would have on the finished car because they simply didn't have the analysis and simulation techniques that we take for granted today; they were only just beginning to understand how useful wind tunnels and model calculations could be.«

Since 1966, the displacement limit for the premier class has been 3 litres for naturally aspirated engines and 1.5 litres for turbocharged engines. Until the advent of turbos (1977 at Renault, and from 1981 for Ferrari), naturally aspirated engines were used by all teams. In its pioneering phase, Formula 1 was divided into two camps: Enzo Ferrari referred to them as the »Grandi Costruttori« or as the »Garagisti«. The league of »grandi costruttori« produce their racing cars – including engines and gearboxes – in-house. In addition to Ferrari, this group also includes Alfa Romeo, BRM, Maserati, Matra and Renault. The Garagisti mostly relied on the Ford Cosworth V8 and were almost without exception based in Great Britain.

However, although this era is considered an ideal playground from today's perspective, a

The 312 B2 is used for the first time at the 1971 Monaco GP, the third race of the season. Jacky Ickx finishes on the podium straight away. The Belgian went on to win the following race. As in Zandvoort, Ferrari also names three cars for the Nürburgring and later Canada: for Ickx, Regazzoni and Andretti

The twelve-cylinder engine with its flat 180° angle gives the 312 B a low centre of gravity

position of responsibility at Ferrari in the early 1970s was anything but an easy ride. The last F1 title dates back to 1964, with Jacky Ickx in particular fuelling the hopes of the Tifosi with individual successes. The pressure is continually high, and the sporting fortunes of the Scuderia are a matter of state importance. And in the meantime, the new shareholder Fiat is also keeping a critical eye on Ferrari. The lack of success is reason enough for Giovanni Agnelli to bang the table in Maranello. In the autumn of 1971, Turin dispatched Fiat engineer Sandro Colombo. His job was to revitalise the racing department, which had fallen behind. He took over the position of chief engineer from Mauro Forghieri. In Colombo's wake, material experts (Fiat Fucine) and aeronautical specialists (Fiat Avio) also came to Maranello, totalling 20 engineers.

The white colour on the exhaust manifold makes any cracks in the spaghetti-like construction immediately visible. Ignition cables and fuel lines can be recognised above it

In 1973, Formula 1 became a non-starter for Ferrari. Jacky Ickx leaves the team in frustration. A new talent, Niki Lauda, joins the team for the following season. The Austrian comes from the British BRM team. »I didn't decide to leave because Ferrari was better back then. Ferrari was also at its lowest point, but offered the conditions for successful work,« recalls Lauda. »I suddenly saw all the wonders of technology, so to speak: my own race track with automatic timekeeping, television monitoring and computers. A comparatively huge factory, lots of mechanics and administrative staff. I couldn't imagine how this team could not be successful.«

»The weak point of the 312 B2 was originally its rear suspension,« recalls Sandro Colombo. The design of the suspension was not compatible with the wide engine. And

The narrow and rustic control centre of the Ferrari 312 B. Jacky Ickx, among others, once sat behind the wheel here when Forghieri tried to compensate for this with low-profile tyres, the vibrations at the rear tended to increase. »Mauro Forghieri did not agree with this diagnosis, however.« So he was sent to the Ferrari research centre in Modena. Colombo put the necessary modifications to the 312 B2 in the hands of Franco Rocchi and racing engineer Giorgio Ferrari.

For economic reasons, Ferrari reduced its line-up to just one F1 car in the course of the 1973 season. A dead end. As a result, Giorgio Ferrari, who had initiated this decision, was transferred to series production. Also, racing director Peter Schetty left the company. As a result Forghieri returned to Maranello and Fiorano respectively. Together with Rocchi and Giancarlo Bussi, he worked on the Ferrari

Fuel level and oil pressure are displayed on the left, rpm in the centre and water temperature on the right. The red pointer is a so-called tell-tale pointer, which stops at the highest displayed value

Discreet power – the combustion air for the 470 hp engine is simply sucked in on both sides of the vehicle in the vicinity of the wheels

The rivets indicate the design of the 312 B2. Ferrari still relies on a tubular frame

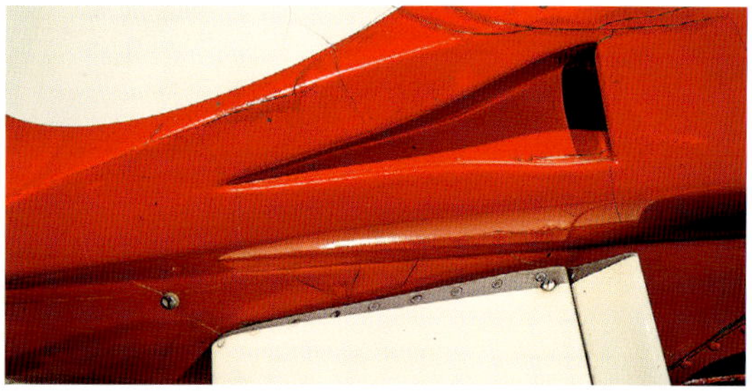

The designers are still experimenting with aerodynamics; scientific meticulousness will come with time

The oil tank is located below the rear wing. Two oil coolers are mounted next to it

312 B3. However, internal disputes continued to thwart the successes that this basically good project could have achieved. The appointment of the 25-year-old Luca Cordero di Montezemolo is another important personnel decision in this turmoil. Enzo Ferrari appointed the motorsport-enthusiastic lawyer as his personal assistant in 1973. He became head of the racing department the very next year. Montezemolo rehabilitated Forghieri, giving him sole responsibility for the 74 model.

Forghieri's father is a Ferrari veteran: he has worked for the company since the 1930s. His son Mauro joined the company in 1959 at the age of 24, with a fresh diploma in his pocket. He was promoted to Sports Director and Chief Designer in 1962 and then to Technical Director in 1969. Despite some back and forth, the Forghieri era at Ferrari, which lasted until 1987, ended with a large number of titles and successes. »He is a technical genius. It's just a shame that he's insane: he snaps, misjudges realities, latches on to them, is not open to arguments and above

At the beginning of the 1970s, Ferrari opted for Firestone tyres. From 1973, Goodyear then supplied the Scuderia with tyres. Firestone has been part of the Japanese Bridgestone Group since 1988

all has the psychological sensitivity of a sand viper,« says Niki Lauda, who nevertheless worked closely and successfully with Forghieri. His predecessor, Jacky Ickx, is not quite so harsh on him: »Mauro Forghieri designed a good car every two years, but instead of tuning it and simply improving what worked well, he was already thinking about the next car.«

As with the mid-engine design, Ferrari was also conservative for a long time when it came to chassis design. As early as 1962, the British Lotus team welded aluminium sheets together to form a lightweight and torsion-resistant chassis for the first time. The monocoque construction method for racing cars was thus invented. Ferrari, meanwhile, remained faithful to the tubular lattice frame with riveted aluminium panels. The 312 B2 was also built using this construction method, which originated in aviation. Its advantage: minor accident damage can be repaired relatively easily.

For the 312 B3, however, Ferrari also opted for a full monocoque made of aluminium sheets riveted together. The design was created in Maranello, but Ferrari enlisted the help of British chassis specialist John Thompson for the realisation. »When the news got out,« recalls Colombo, »there was an outcry.« But it was the right decision. The era was also characterised by great experimentation at Ferrari: »Only a few people know that the B version was developed as a monoposto with two- and four-wheel drive, while the three-litre boxer engine was originally developed for the aircraft sector,« recalls Forghieri, who moved to Lamborghini at the end of the 1980s.

The intake air is fed to the engine in a targeted manner via the periscope-shaped air box. This mobilises additional horsepower

The B in the model designation stands for »Boxer« and thus for the engine design in which the cylinder banks face each other. Opening the bank angle from 60° to 180° lowers the vehicle's centre of gravity by several centimetres. However, as two pistons share a crankpin on the crankshaft in these units, engineers refer to them not as a boxer engine but as a 180° V-engine.

Niki Lauda achieves two victories with the 312 B3-74. Team-mate Clay Reggazzoni brings Ferrari a third. The model was also used in the first two races of the 1975 season. And in the recent past, it is driven once again by ex-Ferrari driver René Arnoux at the Historic GP in Monaco

The era of Niki Lauda's great successes begins with this steering wheel

With the 312 B3-74 designed by Forghieri, Scuderia Ferrari sets the course for further world championship titles

The two orange-coloured snorkels supply cooling air to the internal brakes. The slight transparency of the GRP components is evidence of the engineers' meticulous lightweight construction

Niki Lauda's successful work for Ferrari began with the 312 B3-74. In the end, he has 15 victories to his name

The radiators positioned to the side of the monocoque improve the balance of the vehicle, which now sits on modern tyres

With the step from the 312 B to the 312 B2, the 180° V12 increased from 460 hp at 11,500 rpm to 480 hp at 12,000 rpm. In 1972, drivers were even allowed to rev the twelve-cylinder engine up to 12,600 rpm, with a power output of 490 hp. And in 1974, a further five horsepower was available. This made the Ferrari engine with four valves per combustion chamber and mechanical petrol injection around 20 hp superior to the Ford-Cosworth eight-cylinder engine used by most British racing teams.

A new type of driver enters Formula 1: Niki Lauda, the analytical Austrian nicknamed »the computer« by journalists. His way of working is able to utilise the extensive resources of the »Grandi« racing team and turn them into success.

And not just in the race: Lauda never minces his words and tirelessly drives the team on. He also tests whenever and wherever possible. In doing so, he drives both technical development and the team forward. With the 312 B3-74, Lauda and team-mate Clay Regazzoni achieved a total of three victories in 1974. The Ferrari drivers lead the standings for much of the season. In the end, they were narrowly beaten by McLaren driver Emerson Fittipaldi. But the foundations for the titles in the following years were laid.

FERRARI 312 T

»All of us, technicians, mechanics and drivers, put an incredible amount of enthusiasm and work into this new car, and I have never before or since tested and worked with any car as intensively as with this one,« recalls Niki Lauda of the 312 T, which celebrated its racing debut at the third Grand Prix of the year in South Africa in March 1975.

The »T« stands for »trasversale« and means that the gearbox is located transversely to the direction of travel with the engine positioned longitudinally in the car. This creates space for the rear axle suspension and ensures a more favourable weight distribution. »The conditions were good: everything that was possible was accommodated within the wheelbase,« says Lauda. With around 500 hp, the flat twelve-cylinder engine was considered the most powerful engine in the F1 field at the time.

As in the road car, the drivers sort the five gears through the typical Ferrari gearshift gate. The only difference is that the gear lever is much more delicate and the shift travel is much shorter. The rear brakes are located on the inside of the gearbox, also for better weight distribution. The three gas tanks hold 200 litres of fuel and are located next to the cockpit and behind the driver's seat.

»The car was narrow at the body, the low twelve-cylinder engine allowed every conceivable aerodynamic gimmick for the huge airbox, with optimum airflow to the rear wing. The car was extraordinarily manoeuvrable, perhaps even nervous, but I was able to adapt to it wonderfully.« The car did indeed set new standards, and the Austrian finally brought the world championship crown back to Maranello after eleven lean years.

FERRARI 312 T2

»1976 should have been a splendid season with a secure world championship title,« recalls Niki Lauda, »but the Nürburgring accident happened. Hunt became world champion, I was shattered, Ferrari was demoralised and completely irritated, and technical development came to a complete standstill.« Basically, the 312 T2 is a further development of the previous year's model. In accordance with the regulations, there is now a (rudimentary) crumple zone in front of the driver's feet and a second roll bar at the height of the steering wheel. The rule-makers also reduced the dimensions of the rear wing and banned the airboxes behind the drivers' heads, which had become increasingly larger in the meantime. Instead, Ferrari created air openings in the 312 T2 that run along the side of the cockpit. »The air flows had to be completely re-examined,« says Lauda. »The biggest difficulties were with the airflow to the engine; we had problems with the rooting in the airflow ducts and air reservoirs.« According to the later three-time F1 champion, the car epitomises an exciting era in racing car construction: »Experimenting with aerodynamics, spontaneously and by feel, in a variety of ways, in every nook and cranny. Again and again, holes and scoops and channels – the last years of aerodynamic craftsmanship, before the time when you could still drift in Formula 1 (if you wanted to or had to) came to an end with skirts and underbody mouldings.«

The 312 T2 made its debut at race number two in Spain at the beginning of May 1976 and was used, and subtly modified, in the first two races of 1978. »The 1977 T2 was no longer a superior car. It had the advantage of, let's say, high operational reliability,« recalls Lauda. After all, it was responsible for 24 podium finishes, eight victories, two constructors' titles and one drivers' world championship.

FERRARI 312 T4

In the 1977 season, Lotus brought about a rethink in terms of F1 aerodynamics: the Lotus 78 heralded the era of so-called wing cars, in which the undersides of the car's sidepods were modelled on the wings of aircraft, only in an inverted arrangement. While these generate lift in aeroplanes, they provide downforce in Formula 1, also known as the »ground effect«. At the time, there was talk of 900 kilos. This enables unprecedented cornering speeds that push the drivers to their physical limits – sometimes even beyond them. However, space is needed for the ideal flow around the car. In the Ferrari, however, the wide 180° V12 gets in the way. And yet the engineers in Maranello succeeded in creating a design for the 1979 season that made the red cars more than competitive. Added to this was a stable, powerful engine. Gilles Villeneuve categorised the car for journalists: »Slightly easier to drive than the T3 – one second faster – more grip in the 3rd and 4th gear corners...«.

Once again, Ferrari presented the new model in the middle of the season, at round three in Kyalami, South Africa, at the beginning of March. Villeneuve and Scheckter achieved a one-two victory at the first outing, which the Ferrari drivers were able to repeat at the following race in Long Beach. In total, the Canadian and the South African collected five victories and took a commanding first and second place in the championship – with the better ending for Jody Scheckter, who collected three victories and a number of points finishes. He only failed to score points in three of the 15 races.

With the 312 T5, the 180° V12 underwent further development for the 1980 season. From 1981, Ferrari then opted for a compact V6 turbo, which enabled a consistent wing car design. And so the cards were reshuffled once again...

The successor to the »Queen Mary« has a flatter bonnet than its two-seater sister model, the »Daytona«. Its design embodies the dawn of a new era, and polarised opinion. Even today, the 365 GTC/4 is still considered undervalued

NEWS FROM THE INTERIM PERIOD

In the mid-1960s, resistance breaks out: social movements rebel against the establishment all over the world. Their motivation: cultural reform and to question decades of stifling authority. Many are gathering under the term »1968 movement«, whether in Italy, Germany or the USA. In addition to challenging traditional ways of thinking, it is also about giving justice to the evolving social conditions. Added to this – after a long period of continuous growth – is a significant economic downturn.

In the USA, environmental awareness is also on the rise. In the major cities and conurbations, such as Los Angeles, smog is becoming an issue and cars are being identified as a major polluter. This has resulted in stricter exhaust emission laws. The state of California, one of Ferrari's main sales markets, leads the way. Accordingly, the engineers in Maranello have now also included improvements in emissions output in their specifications. Ferrari clearly felt the importance of the US market and the regulations in force there when the sales figures in the 1968/1969 financial year fell from a total of 729 to just 619 vehicles sold due to severe losses in America.

The 4.4-litre twelve-cylinder engine familiar from the »Daytona« was therefore extensively revised for the 365 GTC/4 presented in 1971, which was clearly aimed at the US market. New intake manifolds mounted on the side of the cylinder heads and six twin carburettors, now arranged horizontally, enabled cleaner mixture preparation. A knock-on effect sees the Colombo V12 now flatter than in the 365 GTB/4, therefore the designers were able to flatten the front end, thus enhancing aesthetics.

Safety is another topic that is moving centre stage. Here, too, the Americans are pioneers. In 1968, the Department of Transportation (DOT) commissioned two companies from outside the industry to design so-called Experimental Safety Vehicles (ESV). Car manufacturers around the world, including Volvo, Mercedes-Benz, Fiat and Volkswagen, picked up the ball and presented their ideas on this topic at the beginning of the 1970s. A striking detail of the vehicles, which were also optimised in terms of crash safety, were deformable, energy-absorbing plastic bumpers. Some of these were soon to be found on production vehicles; the bumpers of Volvo or the Porsche 911 (from model year 1973) are a good example of this. Last but not least, new materials and production techniques are also revolutionising design. This has influenced fashion, architecture and interior design just as much as cars, which are now produced in their multi-millions.

A new elegance – the design of the 365 GTC/4 is a talking point right from the beginning. It breaks with the norm established in Maranello up to that point

Another new feature is that the bumper is part of the bodywork — together with the classic Ferrari radiator grille integrated into it

The instruments also exude the spirit of the coming 1970s. The round gauges now live in angular plastic panels

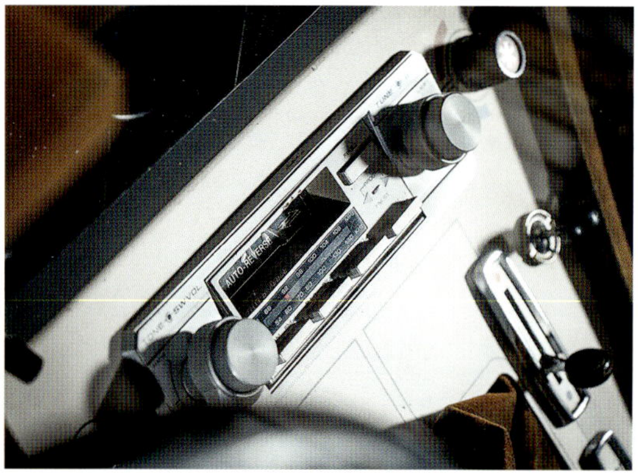

The centre console also has space for a cassette radio and the controls for the standard air conditioning system

The logo shows that the bodywork was created by Pininfarina; the designer was Filippo Sapino

When it came to designing a successor to the 365 GT 2+2, derisively known as the »Queen Mary« due to its opulence, Ferrari and in-house couturier Pininfarina were also on the cusp between the traditional and the new. Ferrari has been 50 per cent owned by Fiat since 1969. The Turin-based conglomerate is responsible for the road-going sports car business and immediately restructures production in Maranello. And an R&D department is set up at Pininfarina, which is soon expanded to include a wind tunnel.

The designers are also engaged in heated discussions about which direction to take. The "Daytona", penned by Leonardo Fioravanti, embodies the classic doctrine of sports car design with its long bonnet, the greenhouse positioned just in front of the rear axle and spoked wheels. On point, the all-round Plexiglas element at the front extends a bridge to modernity. In comparison, the 365 GTC/4 designed by his Pininfarina colleague Filippo Sapino is incomparably more modern: filigree A and C pillars lend the 2+2 with its objectively shaped body a sense of lightness. Instead of chrome trim, black bumpers catch the eye. At the front, the design element incorporates additional headlights, indicators and radiator grille – elements that help form the vehicle's rear end.

This stylistic element anticipated a look that was soon to be seen at all car shows and on the roads. Renault, for example, launched the first large-series car with large-format plastic bumpers in 1972 with the R5. Matra Bagheera, Lancia Beta Montecarlo and Seat 1200 Sport (alias »bocanegra«) are further examples of imitators of this idea.

The first new V12 model since the Ferrari takeover by Fiat is also the first Ferrari to be delivered with folding headlights right from the start. They are also a response to US approval regulations. These stipulate a certain mounting height and now prohibit the use of Plexiglas panelling. Incidentally, »sleep eyes« were first seen on a Ferrari in 1954: the 375 America realised by Pinin Farina on behalf of Roberto Rossellini, which went down in history as the Bergman Ferrari – the Swedish actress Ingrid Bergman was the director's wife.

When the 365 GTC/4 was presented to the press in Geneva in 1971, Filippo Sapino had already been in charge of the design studio of car giant Ford for two years. Therefore he did not have to personally witness how some journalists immediately nicknamed the new design, which broke with traditional design habits, »il Gobbone« (the hunchback). The reason for the nickname is the

The engine compartment of the twelve-cylinder engine is cramped. And thanks to the new cylinder head architecture, the full-size engine is flatter than the »Daytona« offered at the same time

While the air conditioning is part of the standard scope of delivery, the first owner had to order the leather interior separately

The 2+2-seater 365 GTC/4 succeeds the opulent 365 GT 2+2 with the nickname »Queen Mary«. It is known as »il Gobbone«, or »the hunchback«, because of its sweeping waistline

The black plastic front is a new stylistic element that other designers were keen to adopt in the following years. It gives the car a modern look

The four exhaust tailpipes that rise up from the depths of the car's underbody are a visual and acoustic delight. They come from the external supplier Ansa

undulating waistline of the 320 hp sports car; likely though the real rationale is probably its new, unusual appearance.

Unlike the »Daytona« with its transaxle design, the 5-speed gearbox of the GTC/4 is positioned directly on the engine. And instead of the traditional open gearshift gate, the gear lever is in a leather gator. The instruments also speak the language of the 1970s: the round instruments are framed by rectangular plastic elements. While leather upholstery is standard on all Ferrari models, GTC/4 buyers have to order it separately. However, air conditioning, power steering, heated rear window and electric windows are standard, as is a leather steering wheel. The boot and fuel filler flap can be unlocked from the inside and the backrests of the rear seats can be folded down. This creates a second, easily usable luggage compartment, similar to the one that Porsche 911 drivers really appreciate.

While the »Daytona« was produced by Scaglietti, the 500 (other sources speak of 497 or 505) examples of the GTC/4 were made by Pininfarina. At a good DM 75,000, the GTC/4 cost around DM 2,500 less than its two-seater brother, the 365 GTB/4. But while the »Daytona« quickly climbed the Olympus of Ferrari all-time greats, the »Gobbone« long eked out a shadowy existence in the circle of polarising and undervalued Ferraris – similar to the Dino 308 GT 4, the Mondial or the GTC/4 successors 365 GT4 2+2, 400i and 412i.

FERRARI 365 GTB/4 »DAYTONA«

While Lamborghini already favoured the mid-engine concept with the Miura shown in Geneva in 1966, Ferrari once again cultivated the Gran Turismo in the form of a classic front-engined berlinetta. In 1968, the Coupé GTB/4 was presented in Paris, followed by the open-top version in Frankfurt the following year. Incidentally, the 4 stands for the four overhead camshafts of the 4.4-litre V12.

Pininfarina designer Fioravanti brings the Berlinetta into shape. The modern Plexiglas element at the front is at odds with the traditional silhouette and the initially fitted wire wheels. However, the new top model from Maranello manages to impress: the 352 hp Ferrari wins comparison tests with the uncompromising Lamborghini. While the Miura falls 28 km/h short of the top speed of 300 km/h specified by the factory, »auto, motor und sport« measures the »Daytona« at 275 km/h. And the US magazine »Road & Track« summarises succinctly: »The fastest and the best GT is not necessarily the most exotic«.

Both models remained in production for years and underwent various further developments. The »Daytona« was more successful on the market, with 1284 coupés and 125 Spiders being produced by 1973. Lamborghini produces a total of 765 examples of the P 400, P 400 S and P 400 SV series of the Countach's predecessor. Nevertheless, Ferrari also opted for the mid-engine design for its successor, the 365 GT4 BB and 512 BB, which appeared on the market at the end of 1973.

FERRARI 512 BB

Compared to the front-heavy »Daytona«, the steering of the Berlinetta Boxer was »smoother, more direct and more precise«, as racing driver and car tester Paul Frère attested after an initial test drive. »A real improvement«. Maranello spent a long time thinking about the right package for a mid-engined Berlinetta. The developers ultimately decided in favour of a longitudinally mounted 180° V12 with a gearbox behind it. The low-profile engine ensures a low centre of gravity, which is conducive to driving dynamics.

Incidentally, Ferrari likes to refer to the twelve-cylinder engine with the opposing rows of cylinders as a boxer. In view of the arrangement of the connecting rods on the crankshaft, this is technically incorrect. But be that as it may, the design has proven itself in Formula 1 and secured Ferrari a total of seven world championship titles. Frère continues: »The centre of gravity is lower, the weight distribution is more balanced and a modern wheel suspension allows the use of relatively soft springs without compromising road holding on a slippery track. As a result, the car is not only quieter on poor road surfaces, but also safer.«

The story of the BB began at the Turin Motor Show in 1971. Two years later, the first 365 GT/4 BBs, rated by »auto, motor und sport« as having 351 hp, were delivered. In autumn 1976, the mid-engine wedge became the 512 BB, and the conversion to an injection system made it the 512 BBi for the 1982 model year.

FERRARI 550 MARANELLO

With the 550 Maranello, Ferrari returned to the classic transaxle front-engine architecture for its top-of-the-range model. The 485 hp V12 Berlinetta is the first production Ferrari to be presented by Michael Schumacher, who has just signed on with the Scuderia. The press presentation is held at the Nürburgring. »auto, motor und sport« describes the 320 km/h Gran Turismo: »From the headlight domes to the spoiler spoiler edge, the 550 Maranello is a very Italian potpourri of waves, sweeps and curves. From idling speed, the V12 runs in velvet and silk – and when accelerating, the world tips backwards off its hinges. Elegant, quiet and incredibly fast, the 550 whizzes into the rev limiter.« This comes into action at 7600 rpm.

That is enough to open up new dimensions of driving dynamics in terms of comfort and sportiness. In Fiorano, the 550 takes a whole three seconds off its predecessor, the F512 M. Ferrari test driver and Alfa DTM driver Nicola Larini is also at home there. His comment on the manual 6-speed gearbox: »The Maranello actually only needs two gears; the third for starting off and the sixth for all other situations.« In addition to ABS, a two-stage ASR is also on board for the first time to prevent the DM 338,000 super sports car from swerving.

With its magical abbreviation, the 288 GTO references a legendary ancestor. Visually, the two-seater intended for Group B, is close to the 308 - the production numbers champion. Even without a sporting career, it established the hypercar genre

EIGHT CYLINDERS WITH MAGIC POTION

At the end of the 1950s, Albert Uderzo and René Goscinny invented Asterix, a character that almost everyone knows today. The comics centred around the headstrong Gaul, who constantly defies the Romans in conquering the world. They are still successful today: a total of 40 volumes with a total print run of 393 million issues, translated into 108 languages and dialects.

Uderzo, who died in 2020, loved the fine line; first and foremost on paper, but later also on race tracks such as Montlhéry, Nürburgring, Hockenheim and Zolder. Thanks to his success with the little Gaul, he could also afford to call fine automobiles from Maranello his own. At the time, the self-taught driver lived in the outskirts of Paris, in a thatched house that could have come straight out of one of his many Asterix books. Under the roof is a series of Ferraris, both in the form of drawings and in 1:1 scale.

One of these is the 308 GTBi, built in Group 4 trim by Padua-based Ferrari specialist Giuliano Michelotto for French Ferrari importer Charles Pozzi. Beneath its fibreglass bodywork – with parts from the first 308 series called »Vetroresina« – thoroughbred racing technology is used: racing gearbox with limited slip differential and a three-litre naturally aspirated V8 engine producing 300 hp at 8000 rpm. With this 1000-kilo lightweight sports car, European rally champion Jean-Claude Andruet and co-driver Michèle Espinosi-Petit (aka »Biche«) won the Tour de France in 1981 and 1982, among others – a thoroughbred rally event at the time. And her 308 became the tool for the 1981 European Rally Championship runner-up title, with a sister car winning the Italian rally title the

The silhouette of the 288 GTO looks very familiar. It resembles that of the 308 GTB – with a longer wheelbase

MOMO supplies the leather steering wheel. It is still sufficient to be able to steer and honk the horn...

The four headlights for high beam and fog are just as characteristic of the GTO as the raised side mirrors

In addition to the red wrinkle-paint air collectors, the intercoolers dominate the view into the engine compartment. The turbo pressure relief valve can also be spotted

The speedometer scale goes up to 320 km/h. At that time, passers-by could notice this by looking through the side window...

In the 288 GTO, the 5-speed gearbox is also mounted longitudinally. It peeks out cheekily from underneath the rear apron

following year. This lightweight 308, recalls ex-Ferrari sales manager Carlo Tazzioli, motivated Enzo Ferrari to design a new GT sports car. This very vehicle can also be found in Uderzo's garage. At first glance, it looks confusingly similar to the 308; and yet the 288 GTO in the Loh Collection today has virtually nothing in common with it. In fact, apart from their general lines, the two mid-engined coupés only share the windscreen and the shape of the roof and doors.

The 288 GTO was originally designed as a sports car in accordance with the Group B regulations, which were introduced in 1982. From then on, this divided motorsport vehicles into three categories: Group A summarises production or touring cars, Group B henceforth stands for sports cars (produced in small series) and Group C for racing prototypes. While 5000 vehicles produced within one year are required for Group A homologation, 200 are sufficient for Group B. In a way, this category follows in the footsteps of the former Group 4, which previously homologated Grand Tourisme sports cars optimised for racing, of which at least 400 road-legal vehicles must have been produced within 24 months.

The bonnet is made of aluminium and has numerous air vents. It opens to the rear

The rear is impressive – thanks to the spoiler lip, wide tyres and flared wings

The crest usually reserved for racing cars also adorns the 288 GTO, which is intended for the road

The new racing categories prove to be a complete success: Group A experiences its triumphal march with attractive touring car racing series (above all the DTM). Group C ensures full prototype starting fields (in Le Mans, the IMSA series and the World Endurance Championship), and

Group B celebrates its successes in rallying. However, after the premiere of the Audi quattro (1981, still as a Group 4 car), four-wheel drive and (after the appearance of the Peugeot 205 turbo 16 in 1984) compact dimensions are now in demand. In addition, it was no longer possible to win even a flower pot in the World Rally Championship without uncompromising factory commitments. Ferrari is late to the game and remains focussed on the circuit. And while the Reds serve the premier class at the factory, Lancia is responsible for the rally laurels within the Fiat Group.

As a result, there are no sporting commitments for the GTO. Ferrari soon shelved the first considerations regarding the number of units – initially, production of 220 units was envisaged; 200 to fulfil the homologation requirements, a further 20 for the development of a Group B competition version. In the end, 272 units were produced of a vehicle that broadened the horizons of the sports car world in a completely different way. When the 288 GTO was presented in Geneva in 1984, it was the fastest and most powerful road car from Maranello to date. With 400 hp and a weight of just 1160 kilos, a 0-100 km/h sprint in under five seconds and a top speed of more than

Testarossa in the GTO – red accents in the engine compartment

They have long been commonplace in motorsport: three-piece rims

The design once again comes from Pininfarina

300 km/h, it became the first hypercar of the modern automotive era. Other manufacturers followed in its wake, and the supercars presented from time to time also became part of Ferrari's successful business model. Today, the GTO is the rarest of the Ferrari models known as the »Big Five« (including the F40, F50, Enzo and La Ferrari).

Compared to the 308, the GTO has a chassis with an eleven centimetre longer wheelbase. Its body is made of composite materials – as used by Ferrari in Formula 1. The 90° aluminium V8 is mounted longitudinally, has a displacement of 2.8 litres and two turbochargers, including ample intercooling. A twin-disc clutch is responsible for the power transmission between the engine and the 5-speed gearbox protruding boldly from under the rear. »auto, motor und sport« praises the suitability for everyday use of the 110 centimetre flat two-seater, whose 5th gear can be used from 40 km/h without any problems.

As is so often the case, the design is created at Pininfarina, the vehicle in Maranello, and production takes place at the Ferrari subsidiary Scaglietti. Pininfarina designer Leonardo Fioravanti remembers: »No gimmicks were required for the 288 GTO, it had to be simple and clear, uncluttered and clearly a Berlinetta. That was our task.«

The gills in the rear vehicle flanks are a modern interpretation of the styling element typical of the 250 GTO

Enzo Ferrari originally gave the brief of developing a 300 hp sports car. Ferrari engineer Nicola Materazzi replies: »Why only 300? With turbocharging, you can easily achieve 400 hp.«

As has always been customary at Ferrari, the cars are distributed to a select group of customers chosen by Maranello. Albert Uderzo was one of them. The launch of the GTO marked the start of a boom for classic and exotic cars, which would develop into a huge market fuelled by speculators by the early 1990s. Shortly after the premiere of the GTO, which was offered ex works for DM 260,000, market prices rose significantly. For example, when the Porsche 959, offered for DM 420,000, came onto the market in 1987, its purchase price could easily be covered by the proceeds from the sale of a used 288 GTO. Today, the value of the forefather of all hyper-cars is a good two and a half million euros.

The round tail light pairs characterised the Ferrari look for many years. The black Cavallino remains reserved for the 288 GTO

The open gearshift gate is another Ferrari detail

Further slits below the radiator grille are evidence of the air requirement for brake and engine cooling

PORSCHE 959

The FIA Group B regulations came into force in 1982. One and a half years later, Porsche presented its interpretation of the subject at the IAA in September 1983 and anticipated the shape of the sports car that appeared in 1985 as the 959. By the time production began (from 1987), however, Group B was already history – buried after a series of serious accidents involving the rally monsters it had produced up to that point.

»You're always wiser later,« summarised Helmuth Bott, Porsche's Head of Development at the time. »Audi and Lancia, Peugeot and Renault, Leyland, Citroën and Ford entered rallying with their Group B cars – and the Ferrari GTO was neither to be seen on the racetrack nor on the rally tracks.« Both Porsche and Ferrari were wrong about the competitive opportunities for their new sports cars.

The 959 became the first hypercar and the first road-legal model that could compete in Ferrari's league. It is a demonstration of what is technically feasible: electronically controlled all-wheel drive, sequential twin-turbochargers of different sizes, 6-speed gearbox, electronically controlled shock absorbers, hollow moulded rims with air pressure sensors, ABS, aerodynamic bodywork made of composite materials. Plus state-of-the-art engine electronics for perfect response behaviour and optimum boost pressure over a wide rev range.

Before the start of series production, the all-wheel-drive Porsche is used in competition. The version used in the Paris-Dakar desert rally in 1986 proved the qualities of the 959 with an overall victory – even though the 2.85-litre six-cylinder engine derived from the 956/962 Group C racing car only produced 370 hp. The road-going model comes up with 420 hp, and even 450 hp with a performance upgrade offered by the factory. Enough for 315 km/h.

FERRARI TESTAROSSA

The design of the Testarossa is a world sensation when it is launched. The performance of the new top model from Maranello is once again unrivalled worldwide. And despite a top speed of 291 km/h, the designers of this extraordinarily designed sports car dispensed with the spoiler work that had apparently been indispensable until then – the entire vehicle was simply designed in the form of a spoiler. The most beautiful car in the world, and not just for the then GM chief designer Chuck Jordan: countless children's rooms were wallpapered with this dream car, whose body was reminiscent of that of a Group C racing car.

Ferrari no longer wrote the legendary model designation separately. Four valves are now used per combustion chamber under the two valve covers of the 180° V12, which are painted in red crinkle paint. This boosts the further development of the 512-BB engine, which has also been slimmed down by 20 kilos, to 390 hp. As in Formula 1 since the 312 B3, the water coolers of the Testarossa are located in the vehicle flanks. This improves handling and enables aerodynamic design. The shape of the air intakes is also characteristic.

Quite a few loyal Ferrari customers who come away empty-handed when ordering a 288 GTO console themselves with the Testarossa. While the GTO costs DM 260,000 in Germany, a Testarossa costs DM 222,300, with much better availability. The Testarossa is further developed into the 512 TR (428 hp) and F512 M (440 hp), before a front-engined berlinetta once again marks the top of the production model range in 1996 (with the 550 Maranello).

FERRARI MONDIAL T CABRIOLET

Open-top cars had a difficult time in the 1970s. Following safety discussions in the USA, convertibles were now threatened with extinction. »Safety convertibles« with permanently integrated roll bars (such as the Porsche 911 Targa, Triumph Stag or Ferrari 308 GTS) emerged. Rolls-Royce Corniche and Alfa Spider are among the few traditionalists.

But Porsche and Ferrari are heralding the renaissance of open-top motoring. With the Mondial Cabriolet, Ferrari presented its first convertible since the end of production of the 365 GTS/4 in 1983. While only 122 examples of the »Daytona Spider« were produced until 1973, the Mondial opened up a significantly larger customer base. Of the Mondial t alone, the last evolutionary stage of the 2+2, 1017 convertibles (and 858 coupés) were produced. In addition, 629 open-top Mondial quattrovalvole and 810 examples of the Mondial 3.2 Cabriolet are built.

Compared to its 3.2-litre predecessor, the Mondial t has a new vehicle architecture. The 3.4-litre V8 is mounted longitudinally, the gearbox transversely. It anticipated the technology of the 348. Despite its visual proximity to its immediate predecessors, a completely new chassis construction was located beneath the slightly revised bodywork. Equipped with modern engine electronics, the Mondial t delivers 300 hp. That is enough for 255 km/h – open and closed.

The rear view evokes associations with the 288 GTO: the clutch housing peeps out from beneath the twin pairs round rear lights. The Mondial t Cabriolet also experienced the speculation and high-price phase that began parallel to the appearance of the 288 GTO. Although the official price list lists the open-top Mondial t at 200,000 DM, eager customers pay dealers twice as much to get their hands on one.

Its appearance resembles that of a racing car; from the outside, from the inside and when driving. However, it was originally intended exclusively for the road. And because speculators were queuing up at the time, Ferrari increased production figures

THE ESSENCE OF THE BRAND

»The F40 was the most recognisable example of madness. You either walled it up in the basement, or drove it the way it demanded,« muses Gerhard Berger in an interview with »Auto Revue« editor-in-chief, Herbert Völker, about supercars and exotic cars. »I didn't wall it up, and that would have been a very serious mistake.«

With the F40, Ferrari – just three years after the launch of the 288 GTO – went one better. With 478 hp, the uncompromisingly sporty F40 outperforms even the Porsche 959, and while the all-wheel-drive car from Zuffenhausen is bursting with high-tech and comfort, the F40 is the racing car with a road licence: naked, loud, unfiltered.

It is the last car to be presented by Enzo Ferrari. His death, just over a year later, triggers a Ferrari hysteria. In no time at all, prices for classic and new cars soar to astronomical heights. At the time, »manager magazin« wrote: »Bet on red: Ferraris are the best sellers for profit-taking! Don't buy paper, buy steel. Beautiful, red, hand-polished steel...«

Initially, the F40 was listed at DM 444,000. But the number of potential buyers is far greater than the number of vehicles. Ferrari was originally aiming for a production batch of 250 vehicles, recalls Ferrari sales manager Carlo Tazzioli. The rampant speculation is a thorn in his side. Ferrari tried to persuade its customers to hold on to their vehicles for a longer period of time. But quite a few of them are weakening; after all, people interested in the F40 were paying up to six times the list price for a recently owned car on the open market. To counteract this, Ferrari launched further vehicles. The number of units increased to 400, 500 and 750, eventually totalling 1331.

Not only its exterior, but also the far forward seating position evokes associations with Group C racing cars. The design is also state of the art in racing. A steel tubular frame forms the backbone of the vehicle. The body parts are made of composite materials with Kevlar and carbon fibre in a sandwich construction. They are bonded to the frame. Like a racing car, the F40 is almost completely uncovered with the bonnet open. Free access to the technology is obstructed at most by the huge 245/40 and 335/35 x 17-inch rubber tyres. The enormous rear bonnet, with the dome-shaped Plexiglas window, weighs just 22 kilos. No fewer than 37 air openings provide the right amount of fresh air. Lorenzo Ramaciotto, then head of the Pininfarina development centre: »The technicians always want more than the designers.« Incidentally, the bonnet weighed 18 kilograms, the two doors three kilograms each.

The generously dimensioned rear spoiler extends across the entire width of the vehicle. Hot air from the engine, gearbox and turbochargers escapes between the rear lights

The indicator and fog light are concealed under the glass covers. The main headlights are retractable

A steering wheel, pedals with perforated aluminium treads and a felt-covered dashboard – this is what high-performance automotive interior architecture à la Ferrari F40 looks like

A functioning interior air conditioning system is just as important as the correct thermal management of the drivetrain

The classic lettering could be dispensed with – the F40 would always be recognisable as a Ferrari even without it

At least the windows can be opened using a crank handle. The cable pull is used to open the lightweight carbon fibre door

The window in the rear bonnet, which cost DM 100,000 as a spare part at the time, provides a view of two large intercoolers and the three-litre, bi-turbo, eight-cylinder engine underneath. The 90° V8 is basically familiar to the 288 GTO. It was originally commissioned by Fiat's sister company Lancia, where it fuelled the Group C LC2 racer.

Thanks to Ferrari's closeness to motorsport and the purist design of the F40, the hypercar was created in record time: just one year passed between the Supervisory Board's decision and the model launch. This is also helped by the fact that Ferrari has already rolled out a few development vehicles on wheels with the 288 Evoluzione.

Pininfarina succeeds in synthesising racing and road car with the F40

The bucket seats come from the motorsport delivery programme. Kevlar and carbon fibre define the look

Triangular and trapezoidal wishbones ensure that the 1311 Ferrari F40s produced are true to the track

With the bonnets at the front and rear open, the F40 reveals itself like a racing car and provides access to its technology

Shortly before the presentation, Ferrari General Director Giovanni Battista Razelli asks selected journalists for their name suggestions for the new super sports car. Within the company, »Le Mans« was favoured – the successor to the 250 GTO already bore this model designation... Bruno Alfieri, editor of the Italian Automobilia publishing house, suggested naming the vehicle Ferrari Nuvolari. His intention: to name other supercars after other racing heroes such as Ascari, Bandini and Co. Gino Rancati, a friend of Enzo Ferrari, was ultimately awarded the accolade – his suggestion, fitting for the company's 40th anniversary: F40.

»Despite the computer age, we built a car like we used to – with a lot of heart,« says Pininfarina proudly. The F40 marked the pinnacle at the time: no road vehicle at the time could even begin to compare. Even the huge rear spoiler is has no rival. And although the new Ferrari supercar was intended for the road and garage, some examples also enjoyed a successful racing career in GT races, in the BPR series and at Le Mans.

The limited-edition supercars launched in the mid-1980s, such as the 288 GTO, 959 and F40, founded and established the new hypercar category. They can also be described as »instant classics«. »In addition to their brutally eye-catching aesthetics and powerful engines, the Instant Classics justify their price above all with their absolute uniqueness. This makes them an extremely rare phenomena on the road. They are therefore ›virtual‹ cars that are bought not because of their utility, but because of their media significance,« explains design professor Paolo Tumminelli.

Anyone who makes it into the interior of the F40 will see carbon fibre, felt-covered surfaces and perforated aluminium pedals. The hands have a firm grip on the small MOMO suede steering wheel, and the right hand only briefly switches to the spherical gear knob to quickly sort the five gears in the open gearshift gate. The power delivery of the biturbo V8 is overwhelming. »Below 4000 rpm, the power delivery is not yet impressive,« writes »Gazzetta dello Sport« editor Enrico Benzing, »but beyond that, the end of the world breaks loose.« This impresses even seasoned Formula 1 professionals like Gerhard Berger, who experiences a high-speed crash in his F40 while driving along a country road at night. Fortunately, it ended in a meadow. »The damage to the car was ridiculously minor, but I had a doddering knee for two days.«

BUGATTI EB 110

Romano Artioli trades in Ferraris in South Tyrol, imports Japanese cars to Italy and collects old Bugatti cars when Ferruccio Lamborghini tells him about the renaissance of the French luxury brand. Artioli dreamed of the ultimate sports car. And with the EB 110, he realised his dream in 1991. In Campogalliano, not far from Modena, he built a new, ultra-modern factory on the proverbial green field.

The designation EB 110 stands for the 110th birthday of the Italian-born namesake Ettore Bugatti, who would certainly not have regarded the super sports car with disdain. Hidden beneath the bodywork is a 3.5-litre V12 with quad turbochargers and 60 valves, which delivers 560 hp at 8250 rpm. All-wheel drive and ABS ensure grip and driving safety. The sprint from standstill to 100 km/h is completed in 3.2 seconds. The price of the EB 110 is also a superlative: 690,000 Deutschmarks.

From May 1993, the EB 110 Granturismo (the model designation of the »basic« model) was joined by a super sports version with 611 hp and a top speed of over 350 km/h. Romano Artioli's activities put Ferrari under pressure to put more effort into the production cars. Since the beginning of 1993, former Ferrari engineer Mauro Forghieri has been responsible for technology development at Bugatti, as the company has big plans. However, the company is shipwrecked during the economic crisis in the mid-1990s. Production ended in 1995, followed by bankruptcy proceedings in 1997. Meanwhile, Porsche secures the Bugatti name rights. Since 1998, these have belonged to Volkswagen.

A total of 96 EB 110 GTs and 32 EB 110 Supersports were built. Probably the best-known vehicle is the fashionably yellow example with which Michael Schumacher rewards himself for his first F1 title in 1994.

BUGATTI VEYRON 16.4

Ferdinand Karl Piëch, one of Ferdinand Porsche's grandsons, is used to thinking big. He pushed the Porsche racing sports cars from the 906 to the 917 and transformed the brand from class winner to overall winner. He helped make the 911 ready for series production and developed Audi – thanks to quattro, TDI, aerodynamics and aluminium – into a premium brand. As head of Volkswagen, he grew the brand through quality, new models and the acquisition of other manufacturers, such as Bentley and Bugatti.

Piëch has a passion for Le Mans and an aversion to Formula 1: »My Formula 1 is the one-litre car,« says the visionary, who completed his thesis on the Formula 1 engine. Another project close to his heart: Bugatti. Piëch's aim is »to revitalise the brand to the point where it was at its peak in the 1920s and 1930s, i.e. at the top of the automotive world. If anything, the historical values should also be emphasised – in design, in creativity, in elite awareness.«

The result is the Veyron 16.4 with 16 cylinders, four turbochargers, all-wheel drive, 1001 hp, a price tag of one million euros (plus tax) and a top speed of over 406 km/h – the speed at which racing driver Bernd Rosemeyer drove a 16-cylinder Auto Union record-breaking car. For all its performance, the Bugatti shines with the highest standards of luxury and quality as well as everyday driving behaviour.

Over the course of the production period, which lasted until 2015, a large number of special models were created. Bugatti sold the most expensive of these for 2.4 million dollars (plus tax). A total of 450 examples were built, the most powerful of which (Veyron 16.4 Super Sport) produced 1200 hp and could accelerate to speeds of up to 431 km/h – or 267,811 mph that is.

FERRARI SF90 STRADALE

SF90 stands for Scuderia Ferrari 90, which in turn marks the 90th anniversary of the racing team that has characterised the world of motorsport since 1929. Unlike the F1 car of the same name for the 2019 season, the SF90 Stradale is a road-going sports car. It delivers 1000 hp, putting even the La Ferrari hypercar in the shade. The SF90 is also Ferrari's first plug-in hybrid (PHEV), with a 7.9 kWh battery that can be charged via a plug. The battery enables a purely electric range of 25 kilometres. The SF90 Stradale can also be driven exclusively electrically up to 130 kilometres per hour. The hybrid sports car has a top speed of 340 km/h.

The V8 combustion engine with 780 hp is a further development of the engine familiar to the 488 Pista and F8 Tributo. It is supported by three electric drives: two on the front axle and an electric unit connected to the gearbox with a total of 220 hp. Shifting is by means of an 8-speed dual-clutch gearbox. The acceleration capacity of the two-seater is also slightly better than that of the La Ferrari hypercar: 0 to 100 km/h in 2.5 seconds and to 200 km/h in 6.7 seconds. Ferrari (2023) puts the purchase price at 417,890 euros.

Piero Ferrari provides the impetus for the company's 50th anniversary to launch a hypercar with a reference to the first model - open-top and with a twelve-cylinder engine. What is ultimately realised is straight out of Formula 1

FORMULA 1 FOR TWO

Following the success of the F40, the next hypercar was considered early on in Maranello. Enzo Ferrari's second son, Piero Ferrari, comes up with the idea of building a connection to the first Ferrari, the 125 S, to mark the brand's 50th anniversary. The concept of an open-top sports car quickly matures: this time, a twelve-cylinder engine is planned. And like the chassis, the drive technology is also to be derived directly from motorsport. This means that the carbon-fibre monocoque and the V12 borrowed from Formula 1 are a given.

»We have pulled out all the stops with modern automotive technology for the Ferrari F50,« explains Ferrari Head of Development Amadeo Felisa. »It has a carbon fibre chassis, its twelve-cylinder engine is the supporting element of the car's construction, and the springs and dampers work according to the pushrod system.« The F50 is the first road car in which the engine plays a supporting role in the chassis construction – as Ferrari has done since the 1512/158 F1 from 1964. The monocoque offers space for two passengers, behind whom a 105-litre fuel tank is integrated into the structure. As is customary in motor racing, this is a rubber bladder. The chassis with triangular and trapezoidal wishbones and horizontally mounted spring/damper units also comes from motorsport. »The pushrod system enables the springs and dampers to respond very precisely to wheel movements,« explains Giorgio Panini, who is responsible for the development of the chassis and suspension. »Another advantage: greater spatial freedom for designing the axle geometry. The lower links of the rear suspension are positioned high up, which creates space for the diffuser.«

Every single detail is a technical treat. The centrepiece, however, is the engine, which is based on the 1990 F1 power unit. Gerhard Berger likes to think back to the V12s used by Ferrari in Formula 1 up until 1995: »In my memory of fine noises, I revisit the Ferrari twelve-cylinder, which at some engine speeds would get its very own, delicious sound colouring, roaring louder than all the other creatures, and you could even hear the peculiarity of this roar under the helmet and through the earplugs.« In the F50, it will be the most elaborate drive unit Ferrari has ever

The F50 also has large clamshells at the front and rear. The lightweight convertible roof protects against rain, but is not suitable for high speeds

Unlike its predecessor, the chassis of the F50 is made entirely of carbon fibre

The rims of the F50 usually shine in a silvery colour. On the Loh Collection model, they are in the golden colour that Ferrari racing cars have been using on the world's racetracks since the late 1960s

designed for a road car. This includes forged pistons with titanium connecting rods from Ferrari partner, Mahle, and five valves per combustion chamber (three intake valves, two exhaust valves). As with the legendary Dino engine and a number of V12s, the F50 engine also has a cylinder bank angle of 65°. However, while the F1 monoposto has to make do with three and a half litres due to regulations, the road version can draw its power from 4.7 litres of displacement. Consequently, the rated speed – at which the engine delivers its maximum power – is much lower at 8700 rpm than the 15,000 rpm of the F1 engine. As a result, the power output of 520 hp/382 kW is around 200 hp less than what F1 drivers are used to by profession. On the other hand, the naturally aspirated engine, which has been tamed for road use, shines with harmonious drivability, which is similar to that of complex developed production cars.

While Ferrari has been using semi-automatic gearboxes in Formula 1 since 1998, the F50 is shifted manually

As with its predecessor, a grille shapes the air-permeable rear end. Behind the Cavallino is a V12

While electro-hydraulically operated gearboxes had long since established themselves in Formula 1 in the mid-1990s, the F50 still had a manual 6-speed gearbox. The brand-typical open shift gate produces a subtle metallic clunk when changing gear. But even as a manual transmission, the F50 takes 2.5 seconds off its predecessor with a lap time of 1:27.1 on the in-house test track. As with the F40, the F50 also lacks all the little helpers: ABS, brake booster, power steering? Not a thing. However, there is now a digital display – with analogue pointers for the rev counter and speedometer.

The vehicle is presented at the beginning of March 1995. In the presence of Ferrari CEO Luca Montezemolo, Pierro Ferrari, Sergio Pininfarina and Niki Lauda pull the cloth. The guests are then treated to a sleek, Formula 1-inspired design. While the air vents of the F40 still appear to have been carved into the vehicle, they are now harmoniously integrated. This applies to the rear spoiler as well as the radiators positioned horizontally in the front. Modern ellipsoid headlights put an end to the »sleepy eyes« era. The smooth underbody can only be guessed at when looking at the opulent rear diffuser. In Geneva, the vehicle is on display as a Spider. However, the car can be transformed into a coupé in around 30 minutes by fitting the hardtop supplied as standard – provided you have four hands. In the event of unexpected rain showers, the Spider passengers also have access to a tight-fitting emergency soft top. However, this only permits low travelling speeds. Otherwise, the F50 can be travelled at up to 325 km/h – open or closed.

The twelve-cylinder engine does not require a turbocharger. It originates in a straight line from Formula 1 and also serves as the power unit for the 333 SP prototype racing car

On the F50, the Cavallino Rampante is usually bright red in colour. However, some vehicles are also produced in yellow, black, silver or the darker »Barchetta red« colour

Ferrari asked for DM 756,000 from the selected customers who were allocated one of the F50s, which were limited to 349 units. Unlike the F40, they are allowed to choose a paint colour. In addition to the Ferrari red, there is also a choice of yellow, black and the darker »Barchetta Red« and »Nürburgring Silver«. The Loh Collection vehicle is one of two in which the aluminium-magnesium alloy rims have a golden sheen. This colour combination also has its roots in motor racing: Ferrari first used red and gold in the second half of the 1960s.

The valve covers of the F50 are discreetly black. They conceal the drive for five valves per combustion chamber. Titanium connecting rods are among the other engine specialities

The F50 has a wing like its predecessor, albeit more elegantly integrated into the shape of the hypercar

With the open-top F50, Ferrari gives a nod to the 125 S from 1947

FERRARI F1 90 (641)

The 65° V12 with a displacement of 3.5 litres and the semi-automatic transmission, which had also appeared the previous year and now allowed drivers to change gears using paddle shifters on the steering wheel, proved their worth. The further development of the 640, designed by the new Ferrari technical director John Barnard, also proved to be stable. And with Alain Prost, the Scuderia also had a reigning world champion in the cockpit.

The 1990 season was promising: Prost won five Grands Prix, his team-mate Nigel Mansell one race. Once again, it came down to a title fight between McLaren driver Ayrton Senna and his former McLaren team-mate Prost. And once again, the chicane before the start-finish straight is the scene of the title decider: As in the previous year, the two rivals collide, this time with the better end for Senna. For Prost and Ferrari, the title dreams come to an end here.

The V12 was revised once again for the 1990 season, with a shorter stroke, resulting in higher revs (13,800 rpm) and 50 hp more power (710 hp/522 kW). This made it the first naturally aspirated engine to achieve an output in excess of 200 hp per litre. Ferrari had to fight hard for the use of the V12 in the mid-1980s and remained loyal to it until 1995. Up to and including 1994, the displacement remained at 3.5 litres, with another 3.0-litre version being developed in Maranello for the 1995 season.

The electro-hydraulic gearshift was a technology that quickly became a success. After Formula 1, it also conquered other racing categories from 1995 with the DTM Alfa 155 V6 TI from Alfa Corse. From 1998, this technology – which allows the driver to keep both hands on the steering wheel when changing gear at lightning speed – is also available for road vehicles. It celebrated its premiere in the Ferrari F355 F1.

FERRARI 333 SP

With the 333 SP, Ferrari continued its sports car tradition, which had come to an end in 1973 with its withdrawal from the World Sports Car Championship. At the end of 1993, Giampiero Moretti – owner of the MOMO company and enthusiastic gentleman driver – was able to convince Ferrari heir, Piero Ferrari, of his idea of an open-top racing car from Maranello. The car is intended for the new IMSA World Sports Car Championship (WSC) and fills the gap created by the end of Group C and IMSA GTP. As with the 512 PB from 1973, the 333 SP, which was created a good two decades later, also utilised an engine previously developed for Formula 1. While the 65°, twelve-cylinder engine used in Formula 1 until 1994 had a displacement of three litres, the 222 SP had four (and even 4.7 in the F50). It thus mobilises 635 hp/467 kW at 11,500 rpm. By using it in the road-legal F50, Ferrari fulfils the IMSA homologation requirements – and the new five-valve technology is benevolently overlooked by the regulators.

The first vehicles are built directly at Ferrari in Maranello. From chassis number 5 onwards, Ferrari utilises the proven services of Dallara and Michelotto. Ferrari offers the open-top racing sports car to privateers and teams at a price of half a million US dollars. The company charges a further 950,000 US dollars for a comprehensive spare parts package including engine, bodywork, gearbox, suspension parts, brakes and consumables. This means that the 333 SP is no bargain. But for its buyers, it is significantly cheaper than a Porsche 962, for example, which costs 880,000 dollars without spare parts.

MCLAREN F1

The super sports car developed by F1 constructor Gordon Murray and designer Peter Stevens takes the alternative route: it goes from the road to the track. The McLaren F1 with its unusual batwing doors was originally intended exclusively for collectors, road use and as a hypercar benchmark.

As has long been customary in motor racing, a carbon chassis forms the backbone of the vehicle, which offers space for three people in one row of seats – with the driver in the centre. Behind the three seats is a V12 from BMW. The high-torque, extremely well-mannered naturally aspirated engine mobilises 627 hp/461 kW and offers exceptionally smooth running and homogeneous power delivery. Almost every part is painstakingly crafted from the highest quality materials. Of course, this also has its price: 1.5 million marks. McLaren boss Ron Dennis and the Arab entrepreneur (and TAG owner) Mansour Ojjeh, the mentors of this project, also play in this league.

However, banker, amateur racing driver and future Bugatti President Thomas Bscher and Ray Bellm (entrepreneur and racing amateur) approached Ron Dennis independently of each other. They convince the McLaren boss to prepare the F1 for racing – including professional service from the Formula 1 racing team.

The car showed its motorsport genes right from the start and also proved to be a hit on the track. The stage for this was the then popular BPR series for production GT sports cars. He won both titles there, in 1995 and 1996 respectively. In 1995, seven F1 GTRs entered the race at Le Mans. In the end, JJ Lehto, Yannick Dalmas and Masanori Sekiya stood on the top step of the podium. Their marque colleagues Andy Wallace, Derek and Justin Bell finish third. Four McLaren cars are listed in the top five places.

Michael Schumacher and Luca Montezemolo take Ferrari to unprecedented heights. In Formula 1, records are broken. And with its sports cars, Ferrari set benchmarks. The Enzo bears the name of the company founder with pride - and rightly so

FROM THE HANDS OF THE RED BARON

Ferrari wins both Formula 1 titles in 2002. Michael Schumacher was already crowned world champion in Magny-Cours after race 11 of a total of 17, his fifth title. He thus equalled Fangio and broke a record that had seemed unbeatable for decades. Two races later, Ferrari becomes constructors' world champion thanks to the »red goddess«, as the F2002 is known within the team. And just over a month later, Ferrari unveiled a new supercar at the Paris Motor Show: the Ferrari Enzo.

At that time, the F1 teams were still allowed to test extensively. The Ferrari crew made extensive use of the company's own tracks in Fiorano and Mugello. Nobody is busier than Michael Schumacher. And the record-breaking champion never tires of getting involved in the development of the new super sports car. »He started his test drives from the moment we fitted the V12 in a special chassis, right through to the first prototype before the start of series production,« says project manager Giuseppe Petrotta. Schumacher's contributions are extremely valuable for the development work, »especially in terms of electronics, when it comes to the shift strategy, traction control and engine response behaviour,« says Petrotta. But the extraordinary driving ability of the »Red Baron« also opens up new horizons for the developers in terms of braking performance and handling.

The result of these efforts is once again the best-performing road-going Ferrari of all time; the Enzo pulverises the lap time of its predecessor, the F50, by no less than five seconds on the in-house track. »The gap between this car and the Mercedes-McLaren SLR is as big as the gap between Michael Schumacher and David Coulthard. This is now the performance benchmark. This is the king of speed,« judges »Top Gear« journalist Jeremy Clarkson. »EVO editor-in-chief Harry Metcalfe summarises: »Reaching high speeds is incredibly easy in the Enzo. It's as if the car has been coated with a special Teflon coating that allows it to glide effortlessly through the cool night air. In fact, I can't think of any car before the Veyron that builds up speed as easily as the Enzo. You have to be alert though, because at high speeds it feels nervous thanks to the ultra-sharp steering.«

While the Ferrari 288 GTO was still a soloist when it was launched, a number of manufacturers jumped on the lucrative hypercar marketing bandwagon at the time of the Ferrari Enzo: Saleen S7, McLaren F1, Porsche Carrera GT, Koenigsegg CC8S and Mercedes-Benz SLR McLaren are examples of other representatives seeking their fortune in this league.

The 65° V12 has a displacement of 6 litres and mobilises 660 hp/485 kW. The powerful engine is also used in the Maserati MC12 and is utilised by Ferrari (in an enhanced version) to power various models, including the mighty front-

When the Ferrari Enzo is launched, Ferrari can proudly look back on five consecutive constructors' titles. Michael Schumacher has contributed to this as well as to the development of the Enzo

The upward-swinging doors that extend into the roof emphasise the striking appearance

The sweeping diffuser provides plenty of downforce

As in Formula 1, Ferrari also uses the feather-light material, carbon fibre, for engine parts. However, the engineers did not want to omit the shiny silver lettering

The lightweight wheels are mounted on the vehicle with central hub nuts – which, in addition, also provide theft protection

engined Ferrari 812 berlinetta. Shifting is via a sequential 6-speed gearbox. The steering wheel, which is equipped with numerous controls, LEDs and functions, also comes from Formula 1 – even the indicators are controlled via the steering wheel. For the brakes, carbon-ceramic composite discs are used for the first time in a road model. At 92 kilograms, the carbon chassis is another 10 kilograms lighter than the F50 monocoque.

The technical responsibility lies with Amadeo Felisa, General Manager of the Ferrari GT department. He is happy to call in the specialists from Gestione Sportiva, who are responsible for F1 racing, for support and for utilising the F1 wind tunnel that was put into operation in 1998. The Enzo is also characterised by its aerodynamic design, which, thanks to the downforce-generating underbody, manages without a conspicuous rear spoiler. At the front and rear, the Enzo is fitted with spoilers that are harmoniously integrated into the car body and controlled according to speed. The Formula 1-inspired front design and the butterfly doors, which first appeared on the 512 S racing car, are striking.

The design once again comes from Pininfarina. For the first time, it is not an Italian who is responsible, but a Japanese: Kiyoyuki »Ken« Okuyama. The Pininfarina designer's work also includes the Ferrari models 599 GTB Fiorano, California and 612 Scaglietti as well as the Japanese high-speed train Shinkansen E7. He previously worked for General Motors (Pontiac Solstice and Chevrolet Camaro) and Porsche (996 and Boxster). Two decades after the Ferrari Enzo was launched, Ken Okuyama once

The twelve-cylinder engine with a displacement of six litres is good for 660 hp. This enables the Enzo to accelerate to over 350 km/h

The cockpit is easy to enter through the butterfly doors

again found himself in the media in connection with the super sports car: when he was sentenced to four months' imprisonment (suspended) after very clearly exceeding the speed limit (plus 88 km/h when driving on a mountain road with a speed limit of 40 km/h).

Originally, 349 units of the hypercar were planned. However, the discussions held with customers in advance motivated Ferrari to increase the production number to 399. In the end, a 400th model was built in aid of the victims of the tsunami that struck South-East Asia at Christmas 2004. It was presented to Pope John Paul II on 17 January 2005 and auctioned for 1,055,000 euros. The list price for the series models is 662,000 euros; including a few extras, customers usually pay around 700,000 euros.

As is usual in racing cars, the pedals are upright. In the Enzo, there are now only two, as Ferrari now also uses an electro-hydraulically controlled gearstick in its road sports car

The rear grille reveals the racing-style spring/damper units. The rear spoiler, positioned above it, extends depending on the speed

Ferrari increases production from 349 to 399 units – plus one in favour of the tsunami victims

The five double spokes of the wheels conceal brake discs made of carbon-ceramic composite material

For a time, the double lights became a design element of the sports cars from Maranello. Their looks have been familiar for decades

FERRARI FXX

At first glance, the FXX appears to be closely related to the Enzo. In fact, the only thing it shares with the road-legal hypercar is the carbon monocoque. The FXX development begins on the proverbial blank sheet of paper. It is designed as a 1+1-seater.

In fact, unlike the Enzo, the FXX is not suitable for the road. Rather, the selected customers are to contribute to the development of future Ferrari sports cars by passing on their impressions and measurement data directly to the development engineers in Maranello. As with the Corsa Clienti programme, in which Ferrari takes over the servicing and handling of its customers' Formula 1 Ferraris, the FXXs are also taken to Ferrari events selected by the customers. The most prominent of these is the annual season finale »Ferrari in Pista«.

Ferrari originally planned to enter 20 vehicles. However, despite the self-confident price of 1.5 million euros plus tax, production is increased to 29 units. A 30th FXX is being built for Michael Schumacher – intended as a pastime for the record-breaking F1 champion, who will temporarily retire in 2007.

The FXX, which can reach speeds of up to 350 km/h, is 100 kilos (1155 kg) lighter and 150 hp (800 hp/588 kW) more powerful than the Ferrari Enzo. One of the most impressive developments is its aerodynamics. It is closer to Formula 1 than that of road sports cars: while the Enzo generates 500 kilos of downforce, the aerodynamic performance of an F1 car of the time adds up to 1.9 tonnes. The FXX, with all its add-on parts, the aerodynamic underbody and the targeted airflow through the car body, adds up to 1410 kilos. The replacement of the rear-view mirrors with cameras remains an aerodynamic trifle.

FERRARI 812 SUPERFAST

After 6 litres of displacement in the Enzo and 599 and just under 6.3 litres in the FXX and F12 models, the 65° V12 with the designation F140 in the 812 now has 6.5 litres. This boosts the naturally aspirated engine to 800 hp/588 kW (at 8500 rpm), giving it an impressive 123 hp per litre. The maximum torque is 718 Nm, 80 per cent of which is already available at 3500 rpm. This made the large Berlinetta the most powerful front-engined Ferrari of all time in the Ferrari model programme. With a top speed of 340 km/h, a time of 2.9 seconds for the sprint to 100 and 8.3 seconds to 200 km/h, the successor to the F12 Berlinetta and F12 tdf is also the fastest.

The cylinder bank angle of 65°, a Ferrari tradition since the Dino engine, ensures a unique sound thanks to its irregular firing order. The transaxle design, perfected since the »Daytona« and the 550 Maranello, ensures the best possible handling thanks to optimised weight distribution, while the 7-speed dual-clutch gearbox located on the rear axle ensures lightning-fast shifting manoeuvres without interrupting traction. The new electromechanical power steering (EPS) enables precise steering with good feedback and effort adapted to the driving situation. The price for the 812 Superfast is a good 290,000 euros.

In 2019, the coupé will be followed by the open-top version known as the GTS (with hardtop). Another two years later, the 812 Superfast gets an upgrade, 830 hp/610 kW (at 9500 rpm) and the designation 812 Competizione. Typical for Ferrari: this 500,000 euro model, limited to 999 units, is also sold out before it is delivered.

FERRARI DAYTONA SP3

After the Monza SP1 and SP2 models, Ferrari launches a third car in 2021 from its so-called »Icona« series. These small-series super sports cars all quote famous role models. The new Daytona is clearly reminiscent of the 330 P3 and 330 P4 racing sports cars, with which Ferrari achieved an impressive triple victory at the 24 Hours of Daytona in early 1967.

The Monza SP1 and SP2 premiered at the 2018 Paris Motor Show, modelled on the classic 750 and 860 Monza open-top racing sports cars. The modern successors also come without a roof and with a small wind deflector in front of the cockpit. The difference between the SP 1 and SP2, which both reach speeds of over 300 km/h, lies in the number of seats. Technically, the two collector's items, limited to a total of 499 units, are based on the Ferrari 812 presented in 2017, although the 6.5-litre V12 now delivers 810 hp/596 kW.

In November 2021, Ferrari present the now 840 hp/618 kW Daytona SP3 coupé as part of the Ferrari season finale »Ferrari in Pista« on the company's own track in Mugello. Ferrari chief designer Flavio Manzoni and his team have interpreted numerous stylistic elements of the legendary model in a modern way, adapting them to today's size conventions and approval regulations. Ferrari quotes an acceleration time of 2.85 seconds from 0 to 100 km/h, a price of 2,000,000 euros and a production run of 599 units.

2005

2017

2021

La Ferrari – the name embodies the ultimate Ferrari. The 1000 hp sports car is the pinnacle of the extraordinary series of Ferrari hypercars. Consequently, the first hybrid sports car from Maranello marks the dawn of a new era of propulsion

OPUS MAXIMUS OF THE BIG FIVE

With the Ferrari La Ferrari, the company once again demonstrates the pinnacle of what is currently imaginable in terms of road-going sports cars in 2013. As a hybrid, with the combined power of combustion engine and electric drive, the fifth hypercar after the 288 GTO, F40, F50 and Enzo delivers a system output close to the magical 1000 hp threshold. Specifically, the 800 hp/588 kW of the 6.3-litre V12 is complemented by a further 163 electric horsepower or 120 kilowatts. Ferrari has been using hybrid technology under the name HY-KERS (HYbrid – Kinetic Energy Recovery System) in Formula 1 since 2009. It also enables energy recovery. This makes the super sports car not only extremely powerful, but also highly efficient. The combination of both drive systems results in a 40 per cent reduction in fuel consumption and emissions. Unlike later plug-in hybrids, however, the car cannot be driven purely electrically. It takes just 6.9 seconds to sprint from a standstill to 200 km/h, the top speed is 350 km/h and the torque is 900 Nm.

Even Ferrari is at a loss for words – the new super sports car is simply labelled »The Ferrari«, the very essence of the brand. Internally, it is said with a wink that it has become so good »that it has been given a double name.«

The La Ferrari would undoubtedly fit into any art museum. But it also cuts an excellent figure on the road

Ferrari is also breaking new ground in terms of design. Although the La Ferrari has certain elements of in-house models such as the 330 P3, 312 P and the air intakes of the 156 F1, albeit, fashioned in a contemporary way, for the first time in many decades, it is not Pininfarina that is responsible, but the Centro Stile Ferrari, which was founded in 2010. At that time, Ferrari appointed designer Flavio Manzoni as the brand's Head of Design and tasked him with reinventing the brand's identity. »One of the main reasons for founding the Centro Stile was to be able to work closely with the engineers, in other words to combine art and science,« says Manzoni. The technical complexity of La Ferrari is a good example of how closely form and function are ideally linked. The sophisticated aerodynamics include components that are harmoniously integrated into the shape, such as the active rear spoiler, an underbody tail unit and movable elements on the front and rear diffuser.

Flavio Manzoni receives significant support in the design of the La Ferrari from Werner Gruber. Gruber joins from Pininfarina in 2010. He stayed in Maranello for eight years and is now responsible for the design of the US brand Saleen after a stint at BYD. Manzoni shaped his career in the environment of ex-Alfa Romeo and Volkswagen design boss Walter de Silva. At Fiat, he was involved in the Lancia Dialogos, Fulvia Concept and Maserati 3200 GT projects at the beginning of the 2000s, was then responsible for interior design at SEAT and – back at Fiat – designed the Lancia Ypsilon, Fiat Grande Punto and Nuova Fiat 500, among others. At VW, it was then the »de Silva models«

Thanks to numerous additional functions, the steering wheel has long since been expanded into a command centre

The striking rear lights are a contemporary interpretation of the rear lights worn by the Ferraris of the 1960s. The ingenious aerodynamics with spoilers and rear diffuser are a direct result of in-house Formula 1 expertise

from the Golf 6 onwards! He is also responsible for the design reorientation of the Group brands Škoda, Bentley and Bugatti.

The large doors of the La Ferrari, which are reminiscent of butterfly wings, are spectacular. They swing open at the front and open up in an ergonomic fashion into the two-seater cockpit. The steering wheel with its numerous functions immediately catches the eye. It can be used to start the engine, change gears, adjust the shock absorbers and engine characteristics, operate the wipers, lights and indicators. The electronic driving dynamics control is also controlled via the Manettino dial. As in the F1 monoposto, LEDs on the upper edge of the carbon steering wheel provide information about upcoming shift manoeuvres. Behind the steering wheel is a display that can be individually programmed by the driver – just like in Formula 1. Only the display control and functions such as air conditioning, window lifters, hazard warning lights or reverse gear are operated via buttons and controls on the dashboard.

The exterior of La Ferrari is made up of numerous individual elements, which in their combination form a sculptural body

The 6.3-litre twelve-cylinder engine with DSG and hybrid electronics lives under a glass dome

The exhaust air from the front radiator escapes through the opening in the bonnet

The main element of the front spoiler acts like a fin. The doors appear to be wings

Ferrari has always paid loving attention to details, such as this aluminium oil cap

Like the chassis and all body parts, the dashboard is made of artfully crafted carbon fibre. La Ferrari is manufactured at Scaglietti; one body per day, in five stations and eight working hours. The employees of the »Team Serie Speciale« need another 50 hours for the careful assembly up to the final inspection. The number of La Ferrari is limited to 499 coupés – at a list price of 1,190,000 euros each.

Three years after the coupé, Ferrari launched a limited series of 210 units of an open-top sister model. Its designation: Ferrari La Ferrari Aperta. Also in 2016, at the behest of the then Fiat-Chrysler Group CEO Sergio Marchionne, another, 500th coupé was created. It was auctioned off in aid of the victims of an earthquake that struck Italy. The proceeds: 7.5 million US dollars, and the last of the 210 Aperta cars is also donated to charity. In fact, 8.3 million euros were donated to »Save the Children«.

The seats are positioned on the rear wall of the carbon monocoque. The pedals are adjusted to the correct seating position

Usually the name of the first owner can be found in the lower part of the steering wheel

The central element at the rear is reminiscent of an F1 rear wing

MCLAREN P1

McLaren first experimented with hybrid technology – the combination of combustion engine and electric motor(s) – in the 1998 season. »At that time, we were working on a system for recovering and storing the kinetic energy released during braking,« recalls Adrian Newey, who worked there between 1997 and 2005. However, the use of this innovative drive system was prohibited at the time (following protests from Ferrari).

A good ten years later, from 2009, the FIA then authorised the use of a technology known as KERS (Kinetic Energy Recovery System). McLaren (like Ferrari, BMW Sauber and Renault) is one of the four F1 teams that will be using this energy recovery technology in 2009. From 2013, the other competitors followed suit.

Since 2014 – with the introduction of new V6 turbo engines – the use of hybrid technology has been mandatory in Formula 1. At the same time, the LMP1 prototypes entered in Le Mans and the WEC by Audi and Toyota (from 2012) and Porsche (from 2014) will race as hybrids.

Also in 2012, McLaren presented a hypercar with a hybrid drive at the Paris Motor Show – six months before the Ferrari La Ferrari. Its body and combustion engine are based on the MP4-12C. In the P1, the 3.8-litre V8 delivers 737 hp/542 kW. There is also a 132 kW electric motor, giving the plug-in hybrid with a 4.7 kWh battery a total system output of 916 hp/674 kW.

The acceleration figures are: 0-100 km/h in 2.8 seconds, 200 in 6.8 seconds and 300 km/h in 16.5 seconds. The top speed is limited at 350 km/h. The design of the P1 adopts McLaren's characteristic design ethos from 2014. In total McLaren produced 439 units at its modern plant in Woking, UK, at a unit price of 1.1 million euros (plus tax).

PORSCHE 918 SPYDER

Like Ferrari, Porsche is also focussing on hypercars launched in a loose series order. The 959, 911 GT1 and Carrera GT will be followed in 2014 by the first road-legal hybrid sports car from Zuffenhausen. The project was launched in 2009.

As with the Carrera GT, the 918 Spyder also has a carbon monocoque and an engine taken from motorsport: the V8 comes in a straight line from the RS Spyder, which was formerly built for the LMP2 category. What is new is the integration of three electric drives and a 6.8 kWh battery. This gives the hybrid sports car a system output of 887 hp/652 kW. It can be driven fully electrically in city centres, for example, but also shines on the racetrack. On the Nürburgring-Nordschleife, the stopwatch hands stop after 6 minutes and 57 seconds. This means that the all-wheel-drive sports car is over half a minute faster per lap than the rear-wheel-drive Carrera GT.

A total of 918 units are produced in the special series department at the main plant in Zuffenhausen; a few years later, the pre-series models for Porsche's first electric sports car, the Taycan, are built at the same plant.

True to the motto »better is the enemy of good«, Porsche offers a so-called Weissach package for the 918 Spyder hybrid. Surcharge: 71,400 euros. For a total of 839,426 euros, the customer then gets another 41 kilograms lighter model, in which, for example, the wheel bearings are made of ceramic – saving 640 grams in weight. The reward for the effort: from 0 to 100 km/h in 2.6 (instead of 2.8) seconds, to 200 in 7.2 (instead of 7.7) and to 300 km/h in 19.9 (instead of 22.0) seconds. The top speed is 345 km/h.

FERRARI 296 GTB

The 296 GTB was Ferrari's first six-cylinder Berlinetta since the Dino 206/246 GT, which was produced until 1975. In terms of form, the mid-engined two-seater certainly has links to the original Dino, with which the successful history of the »entry-level models« from Maranello began in 1968. The rear element of the roof section and the air vents above the rear wings also evoke memories of the 250 LM. A bit of all right, is how one could summarise the 296.

Hidden beneath the sharp-styled, lightweight bodywork is a three-litre six-cylinder engine that produces 663 hp/488 kW by means of two turbochargers. In addition, there is an electric motor (MGU-K in motorsport jargon) with 122 kW, also acting on the rear axle, which draws its energy from a 7.45 kWh battery housed in the floor of the car. The result is a plug-in hybrid with a system output of 830 hp/610 kW. The special feature of the combustion engine lies in the wide cylinder bank angle: at 120°, it is space-consuming, but – in addition to a more favourable centre of gravity – this offers an ideal installation position for the two turbochargers enthroned above the engine block.

After La Ferrari and the SF90, Ferrari is once again varying the hybrid theme just below the 1000 hp threshold with the 296 GTB. Ferrari quotes a top speed of 330 km/h and acceleration times of 2.9 and 7.3 for the sprint from a standstill to 100 and 200 km/h respectively. The 296 can travel 25 kilometres in purely electric mode and accelerate to 130 km/h. And, depending on the selected driving mode, the electric energy – which is fed into the battery either by recuperation or via the charging socket – is used for electric driving or for boosting. The base price for the 296, which is available as a Berlinetta and Spider, starts at well over a quarter of a million euros.

2013

2014

2022

What was once conceived as an advertising opportunity for the Principality, quickly developed into the highlight of the racing season. The ingredients for for a unique Grand Prix are nowhere richer than in Monte-Carlo

A PRINCELY PLEASURE

»A victory in Monaco has always been a mini world championship title. The race is an important stage in the racing calendar, and if there is one Grand Prix that racing team bosses absolutely have to win, it is this one,« says Alain Prost, describing the importance of the Monaco GP, which has been a permanent fixture on the F1 calendar since 1955. The first Grand Prix Automobile de Monaco dates back to 1929, while the principality on the Riviera first staged the Monte-Carlo Rally in 1911. The city-state sees itself in competition with neighbouring seaside resorts and vies for tourists and attention. In Nice, there is a week of competitions for the relatively new automobile. Accordingly, Monaco is also focussing on the dilettante form of mobility.

Michael Schumacher wins five times in the Principality. The picture shows him on his winning drive in 2001

First of all, however, an automobile club had to be founded and recognised by the global umbrella organisation AIACR (Association Internationale des Automobile Clubs Reconnus – predecessor of the FIA). But after ACM founder Anthony Noghes comes up with a track in the centre of the city and receives support from racing driver Louis Chiron and the Prince, the resistance of the association officials in Paris is broken.

The races with pre-war cars from Bugatti, Alfa Romeo, Mercedes-Benz, Maserati and Delage were already spectacular. The event quickly becomes a magnet for spectators and participants alike. In 1933, Scuderia Ferrari lined up at the start in Monaco for the first time – with four Alfa Romeo 8C 2300 Monza cars. After three and a half hours of racing, Baconin

The photo from 1959 shows the rudimentary safety measures in place at the time – bales of straw had to suffice at the entrance to the casino curve

Borzacchini crossed the finish line in second place. Team-mate Carlo Felice Trossi finished fifth, while Tazio Nuvolari (in the lead after 66 laps) and Eugenio Siena retired. The following year, Scuderia Ferrari's Tipo B cars finished first (Guy Moll), second (Louis Chiron), fourth (Marcel Lehoux) and sixth (Achille Varzi). The Scuderia's dark red was also the dominant colour in 1935, but the team was ultimately beaten by Silver Arrow driver Luigi Fagioli.

The rivalry between Ferrari and the Silver Arrows dominated the scene. A duel that experienced a renaissance at the end of the 1990s. In Monaco, the powerful racing cars

Stirling Moss in the Lotus, the Ferrari 156 F1 with Richie Ginther and Lotus driver Jim Clark form the front row of the grid in 1961. Phil Hill can be seen behind them

have always seemed at odds with their host. But this is one of the key ingredients in the recipe for success of the Monaco GP. »From a rational point of view, no Formula 1 races should actually be held here,« says Alain Prost. This sentence is just as valid today as it was several decades ago. The three-time F1 champion Nelson Piquet described, at the beginning of the 1980s,: »It's like flying through your living room in a helicopter.«

The tightrope act on the narrow strip of tarmac sometimes goes wrong. For example, in 1955, when Alberto Ascari lost control of his Lancia D50 at the entrance to the harbour and flew into the water along with the car. At the last Monaco GP before that, in 1950, he finished second in his Ferrari 125. Behind Alfetta driver Juan Manuel Fangio, who had the presence of mind to take advantage of a pile-up at the start of the race: the Argentinian driver registered the crowd turning away for fear of fire; pre-warned in this way, he was able to pass the accident site marked by leaking petrol and win the race. In 1956, the two Lancia-Ferrari D50s of Collins and Hawthorn collided at the same spot where Ascari had left the track the previous year. This time, the tree trunks installed for safety hold. In 1967, Lorenzo Bandini did not survive a fire accident with his Ferrari at this spot.

Maurice Trintignant did manage a victory in his Ferrari 625 in 1955, but Enzo Ferrari had to wait 20 long years for the next Monaco triumph before Niki Lauda redeemed the Scuderia in 1975. Often enough, the F1 grid in Monte-Carlo shows a balance of power that is independent of the rest of the season. As in roulette, the odds are reallocated in the casino town. However, overtaking manoeuvres are almost impossible and the narrow guardrail channel does not forgive any mistakes. In addition to luck, skill is what counts here. Alain Prost alone won four times, Ayrton Senna even six times. Michael Schumacher brings home five victories in Monaco, three of them with Ferrari (1997, 1999, 2001).

Lauda repeated his previous year's success for the Scuderia in 1976. Jody Scheckter then won in 1979, followed two years later by Gilles Villeneuve. Almost two decades then passed before Michael Schumacher was once again able to represent Ferrari in the royal box. In 2006, he robbed himself and Ferrari of the chance of another Monaco victory with an unfortunate attempt to block his rival Fernando Alonso in qualifying. For the race, he was penalised to the back of the grid and had to let Alonso go in the championship. And so another 16 years passed before Sebastian Vettel secured Ferrari's 10th victory in the Principality in 2017.

The sea, the harbour and the yachts have always given the Monaco GP a unique atmosphere

The 156 F1 passes the old railway station here. The Loews Hotel, built later, has long since given the curve its name

Here the Conte di Trossi speeds through the Principality in 1934 in an Alfa Romeo P3 entered by Scuderia Ferrari

GRAN PREMIO D'ITALIA

The Grand Prix de l'ACF, held in Le Mans in 1906, established the tradition of GP races that continues to this day. In 1921, Italy also gets its Gran Premio, the first race is still held in Montichiari, and from 1922 it is held on the newly built Monza racetrack. In 1924, Antonio Ascari wins with an Alfa Romeo P2 designed by Vittorio Jano. Enzo Ferrari is not entirely innocent in this: he guides the engineer from Fiat to Alfa. In 1925, Alfa Romeo won the first ever GP World Championship title on the spot.

Paul Frère visits Maranello in the spring of 1948. The Belgian recalls: »The Italian Grand Prix was to take place a few weeks later, and Ferrari built its first real monoposto for this race.« Alberto Ascari won Ferrari's first Italian GP in 1949. Since then, there have been 20 Italian F1 victories. And many moments at Ferrari's home races are etched in the memories of the Tifosi. The one-two victory by Berger and Alboreto immediately after the death of Enzo Ferrari in 1988 is one of them, as is the successful return of Niki Lauda after his Nürburgring fire accident in 1976 or Michael Schumacher's tears in the press conference after the race in 2000.

But Monza also means tragedy for Ferrari, such as Wolfgang von Trips' accident in 1961, which serves as the model for the tragic finale of the film »Grand Prix« (1966), in which Ferrari film driver Jean-Pierre Sarti, alias Yves Montand, is thrown out of the car in an accident.

BRITISH GRAND PRIX

Formula 1 was born at Silverstone in 1950. However, the history of the British Grand Prix goes back to 1926. After two years (1926, 1927) at the Brooklands oval near London, the race moved to Donington in 1935 (until 1938). From 1948, the disused Silverstone airbase was used. From the mid-1950s, the GP then alternated between Aintree and Brands Hatch.

Scuderia Ferrari also sent its Alfa Romeos to the island in the pre-war period. From 1949, the Ferrari brand also appeared on the entry lists. What began with two private Ferraris was successfully continued by the factory from 1951: José Froilán González achieved the first Ferrari F1 victory here – over the previously dominant Alfa Romeo »Alfetta« – initiated and built by Ferrari in 1938. In the three years that followed, Ascari and González defended their successes for Ferrari here. Fangio (1956), Collins (1958) and von Trips (1961) continued the triumphs in red. In total, the Scuderia collected 18 GP victories on British soil. Michael Schumacher scored maximum points three times at Silverstone. For Ferrari, the 1999 British GP marked the turning point of a season in which the factory – 20 years after Jody Scheckter's last world championship title – finally has it within its grasp to bring the Constructors' crown back to Maranello. Immediately after the start, Michael Schumacher leaves the track due to a brake failure and breaks his leg when he crashes into the barriers. As a result, Schumacher has to wait another year for the coveted driver's title.

GERMAN GRAND PRIX

In Ferrari history, the German GP is firmly linked with the fire accident that Niki Lauda survived with serious injuries at in 1976. However, it also marked the end of F1 racing on the traditional 22-kilometre circuit.

One of the highlights from a Ferrari perspective was undoubtedly the 1935 Grand Prix, which was won by Scuderia Ferrari driver Tazio Nuvolari ahead of the dominant Silver Arrows on their home turf.

The first Ferrari victory at the German GP dates back to 1950, although at that time the race was still part of Formula 2. 1951 and 1952 saw Ascari repeat his triumph of the previous year. Farina, Fangio, Brooks, Surtees, Ickx, Regazzoni and Alboreto also won for Ferrari at the Ring. Ferrari drivers Lauda, Tambay, Arnoux, Berger, Irvine, Barrichello, Schumacher and Alonso win at Hockenheim. This brings the total number of victories to 20.

However, the history of the German Grand Prix does not begin in the Eifel, but in 1926 on the Berlin AVUS (Automobile Traffic and Practice Road). It then moved to the newly built Nürburgring in 1927. In the 1960s, it was nicknamed the »Green Hell« due to its challenging track layout and the bush hedges close to the track. No other circuit on the F1 calendar is nearly as long as the »Ring«. This is a challenge for safety, but also for television, which is growing in importance. In 1970, the Nürburgring is modernised and Hockenheim is added to the racing calendar for the first time. After the end of the Nordschleife, the Motodrom near Heidelberg became a permanent venue for the German GP. With the opening of the Nürburgring Grand Prix circuit in 1985, there were two suitable GP circuits in Germany.

After five years of hard work, a series of successes and numerous setbacks, Michael Schumacher finally triumphs in red. The title marks the start of a series of five world championships in a row

WORLD CHAMPION AT LAST

»When I was invited to manage Scuderia Ferrari in 1993, I knew that I had a long and hard task ahead of me,« says Jean Todt. »It became even longer and harder.« For seven years, Todt worked as master builder for Ferrari's return to the top of Formula 1, with the 1982, 1983 and 1999 Constructors' titles and the series of individual victories being motivating moments in the long drought leading to a Ferrari driver's world championship title.

Michael Schumacher could have won the title as early as 1999. But even in 1998 and 1997, all options were still open to the Kerpen-based driver in the final. In the end, things turned out differently (broken leg at Silverstone 1999, starting from last position at Suzuka 1998, and a collision with Villeneuve at Jerez 1997). However, in the 2000 season McLaren-Mercedes could once again have claimed the better ending for themselves: Three races before the end of a season it was as exciting as it was balanced, because both Ferrari and McLaren each had seven victories to their name. Ultimately, Schumacher had nine victories, giving him a good lead over the 1998 and 1999 world champion, Mika Häkkinen. At the end of 2000, the knot unravelled and a series of successes began that Formula 1 had never before experienced.

»There were many reasons to celebrate after our fantastic victories, but also disappointments such as our problems in the summer,« recalls the then race director. »Nevertheless, we always believed in ourselves. The team remained calm and, with its cohesion and motivation, proved that it was indeed a ›dream team‹.« The road to becoming a »dream team« was indeed rocky, and not because Todt's first official act was to remove red wine from the racing team's lunch menu. Gerhard Berger experienced the Scuderia as a driver both in the pre-Todt era (1987-1989) and under the motorsport manager who moved from Peugeot to Maranello (1993-1995). »It took until around 1997 for a cool objectivity to prevail in the Ferrari racing team, as would also be the case in English circumstances,« recalls Berger. »The first Ferrari period was still unspoilt by the new practicality of Ross Brawn. At most, there were attempts to get there. But as a driver you want both: The sweet madness that turns us on, and at the same time the clinically sober access to the most mature technology of its time,« says the Austrian about the promising modern Formula 1 standard. However, he also notes: »A Schumacher constellation would have been unthinkable in 1987. The old man would never have allowed the driver to become more dominant than the brand itself. Ferrari was an empire with partly official, partly secret lines of power, and the drivers were simply the outposts that were supposed to do their job: the department to step on the gas.«

Michael Schumacher achieved ten victories in the 2000 season. One of these was in the monoposto with chassis number #203 on display at the Nationales Automuseum

In 1992, Formula 1 boss Bernie Ecclestone is concerned about »his driving force«. He played the headhunter and recommended the little Frenchman to Luca Cordero di Montezemolo, who had been appointed Ferrari boss in 1991 and who could look back on an impressive career from rally co-driver to father of Peugeot's successes in the World Rally Championship, the

The snootiness of Formula 1 at the time — the horizontal, wind penetrating nose is typical of the F1 cars of the time

Relevant references to important technology partners for the team's success – such as Arexons, Brembo, MAHLE and SKF – can be found in the depths of the side pod in front of the rear wheel

Depending on the race track, the rear wing is fitted with different wing profiles, which can also be varied in their angle of attack

The gold-coloured foil protects the carbon chassis parts from the exhaust heat

Situated above the airbox is the aerodynamically shaped mount for the onboard camera – a standard part prescribed by Formula 1

Dakar, Le Mans and the WEC. »Today I know that although Ferrari's comeback lasted even longer than could ever have been foreseen, it clearly began the day a Todt arrived,« says Gerhard Berger.

Another person who initially contributed to the Reds' resurgence was Niki Lauda. The appointment of his former companion was one of Montezemolo's first personalities. Berger categorises it as follows: »That was right in that Lauda didn't care about the Italian cronyism, but spoke plainly and was able to break through the entangled structures.« Christiano Chiavegato, editor of the »Gazzetta dello Sport«, says: »Of course, the hiring of Ross Brawn was just as much a masterstroke as the acquisition of Rory Byrne, both of whom had already worked with Schumacher at Benetton.« In 2000, everything had long since settled into place in Maranello and the Gestione Sportiva cosmos was running like a perfectly adjusted precision clockwork. The customisation of the racing team can also be seen in the fact that around 100 people work in F1 engine development alone under the direction of engine boss Paolo Martinelli and chief engine mechanic Giuseppe D'Agostino. In total, the sports team has 600 employees (out of a total of 1900 at Ferrari).

At Michael Schumacher's workplace, almost all vehicle functions can be controlled from the steering wheel

There are additional switches in the vehicle for reverse gear, radio, rain light and fire extinguisher

The crown of the king – together with Schuberth engineer Oliver Schimpf, Michael Schumacher helps to develop the optimum head protection

The design of the F1-2000 goes back to Ross Brawn and Rory Byrne – the two have already developed Schumacher's World Championship cars together at Benetton

And then, of course, there is the matter of Schumacher himself, who joins Maranello in 1996 as double world champion. This was initially met with incomprehension. ARD described the situation at the time: »It must be the legendary magic of this fiery red toy car that blinds the groom to his better half – Benetton. The diva from Maranello, greying with Formula 1 honours, has been chasing the Miss World title in vain for 16 years, but she has apparently turned dear Schumi's head with her dowry of DM 80 million.« In fact, Schumacher's hard work, skill and meticulousness brought new focus to the team, which, after a series of setbacks in the 2000 season, was finally able to reap the rewards with him.

For the F1-2000, Ferrari has been able to improve the V10, which shines at the right time with the required level of reliability. And when it came to strategy, Ross Brawn's team usually had the better hand. On the track, Michael Schumacher is once again in a class of his own, but still has a highly respectable opponent in Mika Häkkinen. The overtaking manoeuvre in Spa became legendary when Häkkinen and Schumacher passed Ricardo Zonta, who was driving in the middle of the Kemmel straight, on the right and left, with Häkkinen being the first to turn into the next corner. At the finish, Mika and Michael are separated by 1.1 seconds.

With the #203 car on display at the Nationales Automuseum, Schumi wins in Montréal and takes second place in Budapest, behind Häkkinen. At the finale in Suzuka, however, the cards are dealt differently: Schumacher can – thanks to a better strategy – leave Häkkinen behind and secure the world championship crown.

The success serves as an incentive for the »Dream Team«: »I keep telling my employees that we must not forget that we would not be where Ferrari is today without the criticism,« says Luca Montezemolo.

At the heart of the F1-2000 is a three-litre V10, which can be revved up to 17,600 rpm in the race; in qualifying, Ferrari allows its drivers 17,900 rpm. Pneumatically actuated valves, variable-length intake manifolds and an electro-hydraulic accelerator pedal are among the technical insignia of the 102-kilo engine, which delivers 810 hp in the race and 820 hp in qualifying. At that time, the number of engines was not yet limited. Accordingly, they were allocated for qualifying, races, training sessions and tests according to performance. Ferrari used 12 engines per race meeting, resulting in a total of around 110 of these power plants for the 2000 season. There are also eight chassis in total. With #203, Michael Schumacher covered a total of 4436 kilometres during the F1 season.

FERRARI 500 F2

In 1952, just in time for the third season of the Formula 1 story, the cards are reshuffled. Alfa Romeo concentrates on expanding its production capacities and designing new vehicle models. There are no material or human resources left for motorsport. This is a blow to the already meagre motorsport landscape in the early post-war period. The solution of the rule makers: make Formula 2 the first choice.

In 1951, the supercharged Alfa still produced a good 425 hp, while the two-litre naturally aspirated Ferrari produced around 180 hp. Focusing on the weaker vehicles, new competitors quickly emerged. As early as the first race of the season, in Bremgarten, Switzerland, 23 participants were registered. Ferrari competes with a four-cylinder engine. After moderate F1 successes to date, Enzo Ferrari commissioned engineer Lampredi to design a new F1 car. And he was able to convince his boss that the internal friction losses of a four-cylinder engine were significantly lower than those of a twelve-cylinder engine. In fact, the two-litre four-cylinder engine delivered 20 horsepower more than Colombo's two-litre twelve-cylinder engine (166).

Following the withdrawal of Alfa Romeo, Giuseppe Farina (the 1950 F1 champion) joined the Ferrari team – much to the detriment of his father's company (Stabilimenti Farina), whose glamorous period was coming to an end without a new generation of leaders. Ferrari, however, was in a brilliant position: the Scuderia was crowned Formula 1 World Champion in 1952 and 1953 with Alberto Ascari. His car in both seasons: the 500 F2.

FERRARI 156 F1

In 1960, Wolfgang Graf Berghe von Trips gives a foretaste of the cars of the coming season at the wheel of a Ferrari 156 F2 in the 2.5-litre F1 field at the Italian GP. In 1961, Formula 1 once again imposes new engine regulations. And once again, the Formula 2 cars move up a class. Ferrari enters the season well prepared and with a mature 1.5-litre V6, which will end with the drivers' and constructors' titles.

But alongside the joy, Ferrari also had to endure suffering: the darkest moment was the death of the German Ferrari driver in an accident at Monza. A collision with Jim Clark's Lotus cost Wolfgang von Trips and 15 spectators their lives. Trips was well on his way to the F1 title, which then went to his American team-mate Phil Hill.

With the 156, Ferrari also adopted the mid-engine design. The engineers led by chief designer Carlo Chiti also introduced this design to sports cars. The successes in the premier class also silenced the last doubters in Maranello. At the British GP, the Scuderia achieved a one-two-three triumph, and in Belgium all four Ferraris entered finished in the top four. The red dugouts also set themselves apart from the field visually. Characteristic of Ferrari's first mid-engined F1 is a shark's snout separated by a centre bar; hence the racer's nickname – »Squalo«. Ferrari still likes to quote this front design today – as in the eight-cylinder F430 sports car.

FERRARI 1512

The flat twelve-cylinder engine, often referred to as a boxer, celebrates its premiere in the 1512. However, due to the paired connecting rod arrangement on the crankshaft, it was actually a V12 with a 180° bank angle. Ferrari used this architecture in Formula 1 until 1980. Production of the last road sports car with a 180° V12 (F512 M) ended in 1996.

With the flat V12, Ferrari presented the fourth engine concept within four years in 1964 – after the 65° V6 and 120° V6 (1961 and 1962) and the 90° V8 (1963). The 1.5-litre twelve-cylinder monoposto is sometimes referred to as the 512, but in view of the nomenclature typical of Ferrari at the time, 1512 is more appropriate.

With 225 hp at 11,500 rpm, the engine is a powerhouse, its litre output of 150 hp/l is sensational. And it is the first engine to be mounted in the chassis as a load-bearing component. Multiple motorbike world champion John Surtees, Pedro Rodriguez and Lorenzo Bandini sit in the cockpit – with Surtees favouring the more manageable, parallel-mounted eight-cylinder (158). The chassis is built by Fantuzzi using the new »aero« construction method for Ferrari. This involves riveting thin aluminium sheets to the filigree tubular lattice frame. »Our chassis design, adopted from aircraft construction, offered the torsional strength of a monocoque, but cost much less and still allowed repairs to be carried out after minor accidents, as damaged pieces of tubing and sheet metal could be cut out and new elements inserted,« recalls Mauro Forghieri. And that is important, as Ferrari was still a small independent company at the time. The vehicle in the Loh Collection is chassis number #0009, the youngest of the three 1512s produced.

Kimi Räikkönen follows in Schumacher's footsteps; and his debut is brilliant: the Finn brings Ferrari further world championship laurels with his first attempt. So far, 2007 remains the last year in which the Scuderia was successful in creating a world champion.

THE LAST CHAMPION

Despite all the friendly ties, after eleven years Ferrari boss Jean Todt has to install a successor to the throne of the undisputed number one – Michael Schumacher. After five consecutive years in the world championship, the Scuderia saw Renault driver Fernando Alonso snatch the crown from Ferrari twice in a row in 2005 and 2006. His toughest opponent at that time was McLaren driver Kimi Räikkönen. The Finn is ten years younger than the exceptional German driver. After his F1 premiere season with the Swiss Sauber team (2001), he was able to collect numerous victories in five McLaren years. However, despite being spoilt with success, one thing absent for the team from Woking is the world championship title.

Kimi Räikkönen's first race for the red team ends with success. The Fin makes a point and shows that Ferrari is capable of winning even without Schumacher. This silences the critics in Italy for the time being. In his first year at Ferrari, the Finn achieved a total of six victories and twelve podium finishes. Incidentally, his favoured chassis bears the number #262 and is now part of the Loh Collection. Räikkönen drove it in a total of 14 of the 17 races of the season. It has five victories to its name. Ferrari team-mate Felipe Massa – who has been with the team since 2006 – contributed to the constructors' title win with three victories and ten podium finishes.

The Reds are facing sporting headwinds from McLaren-Mercedes, where F1 newcomer Lewis Hamilton and double world champion Fernando Alonso are now successfully in the cockpit. Overall, the Silver Arrows are collecting points diligently. However, McLaren loses important points due to a penalty following a foul by team-mate Alonso against Hamilton, giving Ferrari the second crown. And it gets even worse: at the end of the season, McLaren has to pay the FIA a fine of 100 million US dollars – in comparison: Ferrari operates in 2007 with an estimated budget of 320 million US dollars. The British team was also stripped of all its championship points. The reason for this is an espionage affair: McLaren obtained data from Ferrari's long-time chief mechanic Nigel Stepney.

Eleven years after Schumacher's Ferrari debut, the »Dream Team« is characterised by major upheavals. Räikkönen's signing was already initiated in 2005, but always denied until »Schumi's« retirement in 2006.

Jean Todt remains team boss of the Scuderia until 2008. Between 2006 and 2008, he was also Ferrari CEO before being elected President of the FIA from 2009 until the end of 2021. Todt hands over the position of Team Principal to Stefano Domenicali at the end of 2007. The current CEO of Formula 1 (from 2021) remained in this role until after the disappointing start to the 2014 season. His successors are ex-Ferrari North America boss Marco Mattiacci (2014 season), ex-Marlboro manager Maurizio Arrivabene (late 2014 to early 2019), Ferrari F1 engineer and ex-technical director Mattia Binotto (2019-2022) and ex-Sauber team boss Frédéric Vasseur (since 2023).

Ross Brawn is also looking for new challenges after his success as Technical Director of Ferrari at the end of the 2006 season. He became team principal at the Honda Formula 1 team

Räikkönen wins five times with the Loh Collection F2007. At the end of the season, he puts his rival Lewis Hamilton in his place with 110 to 109 world championship points

The protective element made of mouldable foam on the edge of the cockpit bears the autograph of Kimi Räikkönen

The F2007 vividly illustrates the effort that the teams put into aerodynamics at that time

The chassis parts also have an aerodynamic function

that had emerged from BAR and achieved world championship honours in 2009 – now under his own name – with Jenson Button, before selling his Brawn GP team to Daimler after just one year. He remained Team Principal of Mercedes Grand Prix until the end of 2013 and has been Sports Director of Formula 1 since 2017. F1 constructor Rory Byrne – also part of the Ferrari »Dream Team« – followed Schumacher from Benetton to Ferrari in 1996. There he replaced John Barnard and created a series of winning cars. The last F1 Ferrari to bear his signature was the F2004, where he also retired from day-to-day business after Schumi's departure from Ferrari F1, but still supports the team as a consultant to this day. Meanwhile, his successors Aldo Costa (until 2011), Nikolas Tombazis (until 2014), Simone Resta (until 2020) and Enrico Cardile (since 2020) take over the day-to-day business.

The driver line-up in Maranello is also in disarray. Eddie Irvine (1996-1999) and Rubens Barrichello (2000-2005) drive alongside Schumacher. Felipe Massa arrives in 2006. He stays until 2013 and Kimi Räikkönen initially signs with Ferrari for three years. In 2010, he makes way for Fernando Alonso. Four years later, the Finn takes

The barcode allows sponsor Marlboro to circumvent the restrictions on cigarette advertising in the sports environment

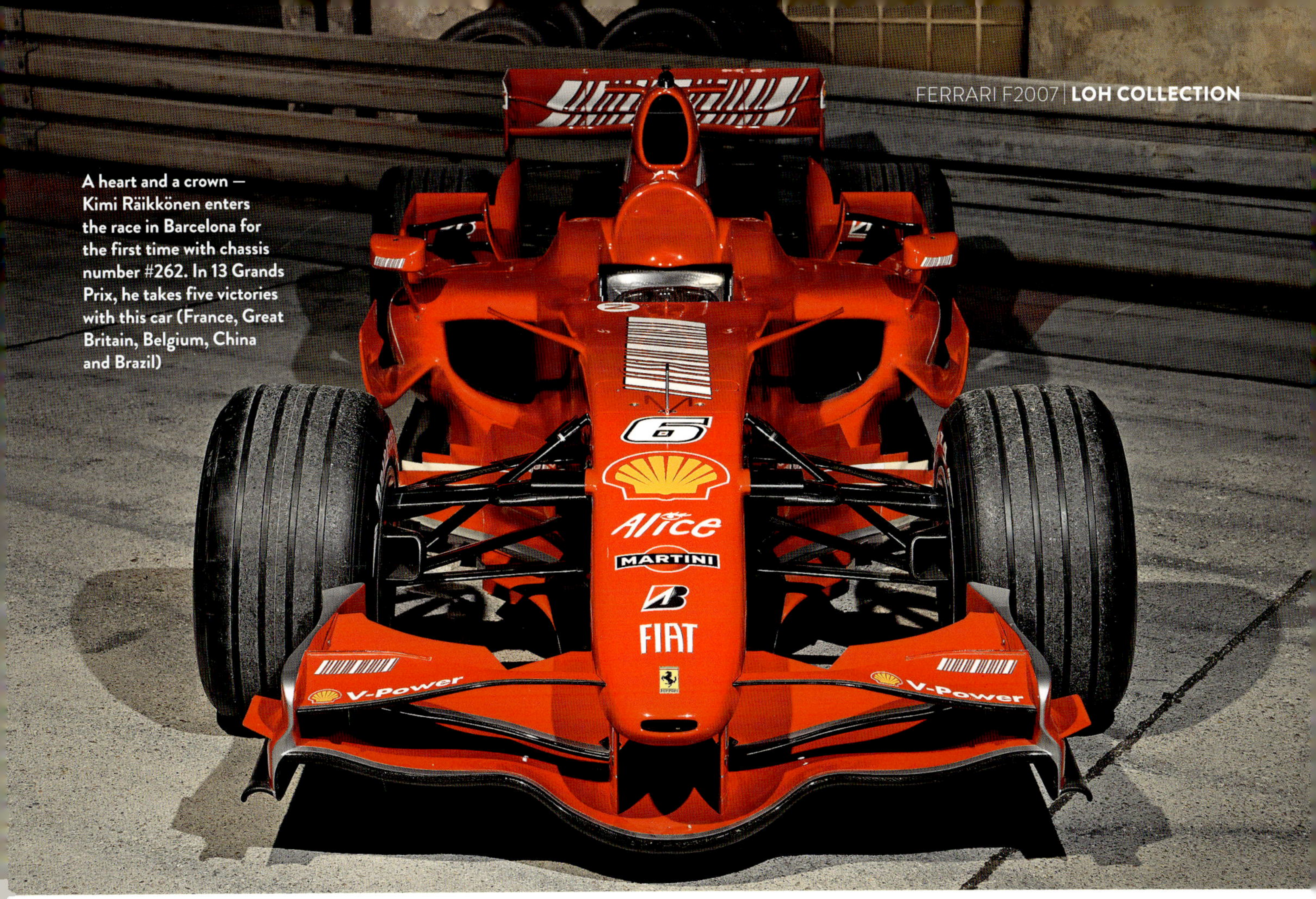

A heart and a crown — Kimi Räikkönen enters the race in Barcelona for the first time with chassis number #262. In 13 Grands Prix, he takes five victories with this car (France, Great Britain, Belgium, China and Brazil)

over Massa's cockpit on his return to Ferrari. Räikkönen remains with the Scuderia up to and including 2018 and then switches to the Ferrari customer team Alfa Romeo (Sauber). Alonso drives for Ferrari for five seasons until 2014. The Spaniard achieved the runner-up title three times with the Reds. Sebastian Vettel takes his place from 2015. The Heppenheim native emulates his idol Michael Schumacher and has four world championship titles (with Red Bull) under his belt. In six seasons (up to and including 2020), he racked up 14 victories, but did not bring the crown back to Maranello. In Charles Leclerc, he had a hungry team-mate in 2019. When the team temporarily favoured the Monegasque, Vettel packed his bags. Carlos Sainz becomes a Ferrari F1 driver in 2021.

In the 2007 season, standard tyres are used for all teams for the first time since 2000. The supplier is long-standing Ferrari partner Bridgestone. »A constant issue in the 2007 season was the difficulty of overtaking,« recalls F1 designer Adrian Newey. »The cars were so aerodynamically optimised that position shifts often only came about through clever pit stop strategies. An ›Overtaking‹ working group was set up to work out changes to the aero rules so that the cars would be less hindered by the wake tur-

Another spoiler sits enthroned above the exhaust air openings in front of the rear axle

Numerous components illustrate the meticulousness with which Formula 1 engineers optimise the aerodynamic shape of their cars

Räikkönen secures the title with a brilliant final spurt in a thrilling season. In the Constructors' Championship, the F2007 is clearly superior; thanks to the additional successes of team-mate Felipe Massa

Räikkönen completes more test kilometres than any of his competitors in 2007. The ingenious shape of the numerous aerodynamic parts is calculated by computer, tested in the wind tunnel and trialled on the track

The wheel covers are an important detail for the optimum temperature balance of the tyres. Their spoke look is merely glued on – to the delight of development partner BBS

bulence of the car in front.« The Ferrari F2007 is regarded as the most aerodynamically efficient car in the Formula 1 field. It is the second car from the pen of Aldo Costa and has a longer wheelbase than the cars of the competition. This gives it the largest surface area. And this can be utilised for aerodynamics. The numerous wings, flics, flaps and bargeboards are the result of extensive wind tunnel work. Clever weight distribution and electronic differentials make up for the disadvantages associated with the wheelbase.

Although Räikkönen clocks up more test kilometres than any other F1 driver in 2007, reliability remains the Achilles heel of the F2007: Räikkönen retires in Barcelona and at the Nürburgring, Massa even drops out three times – and thus out of the title race. Nevertheless, Räikkönen manages the feat of winning the world championship crown in his debut season. Although Kimi had to accept a 26-point deficit during the year, he delivered what Jean Todt had bought him for. He praises him accordingly: »He is modest, determined and not as emotionless as many people think, you just have to translate his feelings correctly.«

In the following year, Ferrari built on the previous year's victories; now with an advantage for Massa. The Brazilian won five Grands Prix, while his Finnish team-mate only managed to finish first twice. His toughest rival is McLaren driver Lewis Hamilton. At the finale in São Paulo, the curious situation arises in which Massa is allowed to feel like the world champion for a few seconds. He wins the race and is also ahead in the championship. However, due to shifts in the midfield, Hamilton manages to snatch fifth place on the last corner and becomes champion.

FERRARI LANCIA D50

»The only car that could have stood up to the Silver Arrows was the Lancia D50 designed by Vittorio Jano, the Alfa Romeo designer of the heyday from 1924 to 1937,« says Paul Frère. »It was very, very fast, handled well and only lacked the necessary reliability, which could only have been achieved through appropriate development work. But Lancia ran out of money, and at the end of 1955 the entire Lancia racing team was handed over to Ferrari.« The financial bottleneck weighed just as heavily in Turin as the loss of Alberto Ascari, who was killed in a test drive with a private Ferrari in Monza in 1955.

The Ferrari 625 F1 and 555 F1 »Super Squalo« models of the time were clearly inferior to the Jano design. The fuel tanks on the D50s were positioned laterally between the wheels to ensure harmonious weight distribution, regardless of the fuel volume carried. What cannot be seen at first glance: hidden under the bodywork is a drive train mounted at an angle to the longitudinal direction. It enables a particularly flat cockpit.

The D50 was revised for the 1956 season and the Argentinian Juan Manuel Fangio found his way into the Ferrari cockpit. The team achieved a total of five victories and ten podium finishes. In 1956, a GP race distance was still around 500 kilometres and the drivers of a team were allowed to swap cars. As a result, six more Ferrari drivers – Peter Collins, Eugenio Castellotti, Paul Frère, Luigi Musso, Alfonso de Portago and Olivier Gendebien – finished in the points in the F1 standings. And Fangio wins his fourth of five world championship titles.

»The car converted by Ferrari was certainly no better and was heavier than the original Lancia,« summarised Frère, »but at least enough money had been invested in its development to make it reliable, and it was undoubtedly the best car in Formula 1 in 1956.«

FERRARI 126 C2

With the second turbo model, the Scuderia had the technology completely under control and the revised chassis in monocoque construction proved to be rigid. In addition, the Reds were now running on Goodyear tyres instead of Michelin. The season could have been a runaway success for Ferrari, but 1982 actually turned out to be a sad year in which even the Constructors' World Championship title was of little consolation.

The aerodynamicists in the team designed the new car entirely for »ground effect«. The slim V6 turbo with a 120° bank angle makes this possible. The downforce is so high that the Ferrari drivers sometimes even do without a front spoiler. The 1.5-litre engine mobilises 580 hp in racing mode, and there are even rumours of over 600 hp – thanks to water injection. The new mastermind in the team is Harvey Postlethwaite, who comes from Wolf Racing and Fittipaldi.

However, the season has been characterised by disputes between Bernie Ecclestone, the representative of the British racing teams known as »garagisti«, and FISA President Jean-Marie Balestre, representing the works teams. There is another internal dispute within Ferrari: Gilles Villeneuve and Didier Pironi. This rivalry has Villeneuve going to extremes. In qualifying for the Belgian GP, he paid for his efforts with his life. Pironi also had an accident during the course of the season. Patrick Tambay and Mario Andretti were also added to the C2 cockpit.

In the end, three victories and eleven podium finishes are in the Ferrari books, but the title goes to Williams driver Keke Rosberg, who is the most conscientious in scoring points (and a victory).

FERRARI F1-87/88 C

On the 90th anniversary of the Commendatore, Ferrari starts the new racing season with optimism. At the end of the 1987 F1 year, Ferrari driver Gerhard Berger scored two consecutive victories. They fuelled hopes of a good year and a comeback to the top of the standings. But 1988 was also a year of transition: it was the last season with turbo engines for the time being – they would not return until 2014 in the form of 1.6-litre V6 engines combined with hybrid technology.

A maximum of 150 litres of fuel is available for the 300 km race distance and the boost pressure is limited to 2.5 bar. This gives the 1.5-litre V6 from Ferrari between 620 and 650 hp at 12,000 rpm. However, the chassis was merely a further development of the previous year's car, because behind the scenes the new chief designer John Barnard (the »father« of the first carbon monoposto McLaren MP4), who had been poached from McLaren, had already been working flat out on the new car for the 1989 season since the previous year.

For Gerhard Berger and Michele Alboreto, it was once again a mixed season. »Our big problem was simply the powerful Honda V6 engine,« says Ferrari engine man Enzo Corradini. Gerhard Berger speaks of »fuel misery«.

15 of the 16 Grands Prix went to McLaren, ten of which were double victories. At least Berger and Alboreto achieved a Ferrari one-two at the Italian GP in Monza – the Scuderia dedicated it to the company founder who had died shortly before on 14 August 1988. At the end of the season, the two Ferrari drivers finished third and fifth behind the dominant McLaren duo of Senna and Prost thanks to eight podium finishes.

The fascination with Ferrari cannot be described by numbers and figures alone. But you can try...

INTERESTING FIGURES

243

F1 WORLD CHAMPIONSHIP VICTORIES
(as of 2023)

15

F1 DRIVERS' WORLD CHAMPIONSHIP TITLES
(1952, 1953, 1956, 1958, 1961, 1964, 1975, 1977, 1979, 2000, 2001, 2002, 2003, 2004, 2007)

Enzo Ferrari is
21 years
old when when he competes in his first car race.

10

overall victories in the 24-hour race at Le Mans

16

F1 CONSTRUCTORS' WORLD CHAMPIONSHIP TITLES
(announced since 1958)
(1961, 1964, 1975, 1976, 1977, 1979, 1982, 1983, 1999, 2000, 2001, 2002, 2003, 2004, 2007, 2008)

13.221 ***VEHICLES SOLD IN 2022.***

7

FIA WEC GT World Championship titles
(announced since 2012) (2012, 2013, 2014, 2016, 2017, 2021, 2022)

7

overall victories at the Targa Florio
(1932, 1933, 1934, 1935 – Scuderia Ferrari on Alfa Romeo plus 1948, 1949, 1972)

13

overall victories at the Mille Miglia
(1933, 1934, 1935, 1936, 1937 — Scuderia Ferrari on Alfa Romeo plus 1938 as Alfa team principal and for the Ferrari brand 1948, 1949, 1950, 1952, 1953, 1956, 1957)

In 2023, Ferrari employed just under

5000 PEOPLE

Ferrari's stock market value in August 2023 is

55.8 billion

US dollars.

13

WORLD SPORTS CAR CHAMPIONSHIP TITLES
(CONTESTED BETWEEN 1953 AND 1992) (1953, 1954, 1956, 1957, 1958, 1960, 1961, 1962, 1963, 1964, 1965, 1967, 1972)

Ferrari received

7 billion

lire, the equivalent of **1.2 MILLION** US dollars, from Fiat in 1969 for 40 per cent of its shares in the company. Fiat had been supporting the racing and sports car manufacturer since 1966.

444.000

LIST PRICE IN DM FOR THE FERRARI F40
at its market launch in 1989

3

FIA GT TITLES
(contested between 1997 and 2009) (2006, 2008, 2009)

In 2015, Fiat raised

893 million

US dollars by selling ten per cent of its shares on Wall Street.

18 million

US dollars was to be the price for the sale of the company to Ford in 1963.

20-25 million

US dollars were offered by Ferrari customer and oil entrepreneur Mecom for the company in 1962.

The Nationales Automuseum has spontaneously established itself as the premiere address for living automotive history. In addition, the Loh Collection is also a satellite campus of various universities. Last, but not least, the FIA has given it a special award...

AN EXCELLENT EXPERIENCE

The Bucciali is one of the impressive curiosities in the Nationales Automuseum. The opulent bodywork of this rare vehicle conceals a front-wheel drive system

The FIA recently awarded the Nationales Automuseum the »Founding Members' Club Heritage Cup« – a prestigious honour. Tobias Reichle and Florian Urbitsch (right) accept the award at the FIA awards ceremony at the end of the 2023 season

The Nationales Automuseum– The Loh Collection has taken the automotive, art and technology culture world by storm and has established itself as one of the world's leading automobile museums. The FIA honoured the Nationales Automuseum with the Heritage Cup of the »FIA Founding Members' Club« (FMC) at its annual winners' gala. This award is regarded as an accolade in the classic car scene, as only select events, organisations or museums are honoured in this way. This means that the Nationales Automuseum is now mentioned in the same breath as the Mille Miglia, the AvD Oldtimer Grand Prix, the Louwman Museum in The Hague, the Rolex Monterey Motorsports Reunion or the London to Brighton Veteran Car Run.

In the first three months since the exhibition opened at the end of July 2023, more than 30,000 visitors have already visited the Loh Collection in Dietzhölztal-Ewersbach. They

In the 1990s, Ellen Lohr shook up the DTM starting field, dominated by male competitors, with this Mercedes 190. With the »Berlin 2000« livery, Mercedes supported the Olympic bid of Germany's new (and old) capital after reunification

Mika Häkkinen's McLaren-Mercedes is the successful counterpart to Schumacher's Ferrari at the end of the 1990s

The Nationales Automuseumknows how to fill the historic industrial site with life in a unique way

The Maybach Exelero (right) is a spectacular one-off and is the result of an initiative by tyre manufacturer Fulda

experience a unique selection of special vehicles that bring the history of the automobile to life. Whether super sports car, pre-war saloon, economic miracle car, Formula 1 racer, or racing prototype – every car has its own story to tell. »That's what makes the Nationales Automuseum so unique,« explains museum curator Tobias Reichle. »The selection is so diverse that everyone will find their own favourite.« Among the most admired exhibits are the Maybach Exelero (also known from Jay-Z's music videos) initiated by tyre manufacturer Fulda and the French Bucciali. »This super rare luxury vehicle from the pre-war era has only been seen in person by very few people before.«

The ambience of the museum, which has found its home in a historic industrial ensemble, is also unique. The exhibition designers know how to make ideal use of the

Among the DTM legends on display is the Alfa Romeo 155 V6 TI, with which the Italians dominated the world-renowned touring car series in 1993 and 1994

The exhibition area, modelled on a steep curve, dominates the large exhibition hall. But there is also spectacular eye candy at its backside

The Mercedes CLK-GTR and Bugatti Veyron can be seen in the foreground, with the graffiti-decorated industrial architecture in the background

different rooms. In the large main hall, there is space for a large steep curve on which various vehicles can be seen. The backdrop is formed by an oversized box, into which the exhibits are transported by a kind of paternoster. Elsewhere, the vehicles stand in front of a street façade with an accompanying light show theatre. The special exhibition »100 Years of Le Mans« is now followed by a unique exhibition on the subject of Ferrari. »There has never been an exhibition of this variety of important vehicles in this form before,« says Florian Urbitsch, Executive Director Marketing & PR of the Loh Collection. »And in addition to the jewels for the racetrack and road shown in the rooms of the special exhibition, there are myriad other Ferraris to be found in the museum.« These include the Ferrari racing transporter from 1967 and a Peugeot 404 Familiale, which the Ferrari racing team also used as a service vehicle in the 1960s.

The Nationales Automuseum has also long since established itself as a focal point for brand clubs and classic car rallies. The Siegerland-Classic-Rallye has already integrated the museum, which is located in the Lahn-Dill district, into its route, as have numerous clubs that stop off there. The on-site catering facilities cater for the physical well-being of visitors. The museum organisers have even thought of camper van pitches with electricity and water.

Other crowd-pullers are the in-house history workshop and the museum's own shop. In addition to books and other automobilia, there are also model cars in various scales specially produced for the Nationales Automuseum. Among the best sellers are special models of exhibition vehicles in 1:87 scale, but limited edition rarities in 1:18 scale also sell out quickly.
With the launch of the Nationales Automuseum – The Loh Collection, Nürtingen-Geislingen University of Applied Sciences (HfWU), under the direction of Dipl.-Ing. Prof. Dr Jochen Buck, has opened a university campus here far from home since the winter semester of 2023. Thanks to the cooperation with design from professor Wolfgang Henseler of Pforzheim, Dietzhölztal-Ewersbach is now also a science centre for students of automotive design.

The Nationales Automuseum also provides unusual insights and thus makes technology tangible

Niki Lauda, Stefan Bellof, Ayrton Senna and Mika Häkkinen are the names of the drivers of these four Formula 1 cars from three decades

The BMW M3 is the world's most successful racing touring car in motorsport history. The Group A racers have long since earned a permanent place in the eternal Hall of Fame

An excellent museum

The International Automobile Federation (FIA) recently honoured the Nationales Automuseum. It is a very special place. This also applies to the inner courtyard. Particularly when very special vehicles come together here. In the foreground you can see the »Ferrari Big Five«, in the background two exclusive design gems and the competition versions of the 212 and 250 GT

ABOUT THE AUTHOR

Jörg Walz is now the author of two dozen books on various automotive and motorsport topics. The motoring journalist has numerous experiences with Ferrari. Starting with a trip from his home town of Lübeck to Ferrari importer Auto Becker in Düsseldorf. The requested Ferrari brochures are still part of his library of around 3000 motoring titles today.

Walz first came into close contact with the Ferrari F40 (as well as the Porsche 959 and Ruf CTR) during a holiday job at »auto, motor und sport«. As editor-in-chief of the Italian magazine »Volante«, the author was then allowed to drive the then dream quartet 250 GTO, 288 GTO, F40 and F50 around the Nürburgring-Nordschleife himself. Numerous Ferrari experiences followed; for example during his Formula 1 time with RTL and AUTO ZEITUNG or at Ferrari's 60th birthday in Maranello – including travelling there in his own Ferrari...

After holding various positions in the automotive industry, Jörg Walz now works in marketing at MAHLE, another long-standing partner of the Cavallino Rampante.

IMPRINT

Publisher:
Nationales Automuseum – The Loh Collection, Dietzhölztal-Ewersbach

Book idea:
Prof. Dr. Ing. E.h. Friedhelm Loh

Conzept, layout, typesetting and words:
Florian Urbitsch (Nationales Automuseum), Jörg Walz (conzept, words)
Markus Köllmann and Lukas Kretzschmar, tebitron Gmbh, Gerlingen (layout, typesetting)
Georg Otto, Joachim Kuch, Joachim Hack
(project management, editing and final editing)
Matt Keene (English editing)

Printed in Czech Republic

Images and illustrations:
Nationales Automuseum, Dominic Fraser, Burkhard Werner, ACO, Alfa Romeo, ATP, Ferrari, Ford, Getty, HochZwei, Louis Klemantaski, Marlboro/Philip Morris, McLaren, Porsche, Stellantis, Jörg Walz; Ferrari: 35, 37, 38, 39, 84, 87, 127, 129, 159, 171; Getty Images: 1–6, 15, 17, 21–29, 32–37, 40–45, 47, 49, 50, 54–57, 63, 69, 77, 85–88, 95–99, 107, 119, 121, 127, 129, 130, 136, 139, 159, 165, 180–183, 184, 189, 190, 195; picture alliance: 33, 36, 38, 63, 83, 99; shutterstock: 34, 69, 113, 119, 165, 145, 172.

Thanks to:
Prof. Dr. Ing. E.h. Friedhelm Loh, Tobias Reichle, Florian Urbitsch, Uwe Schüler, Lars Werner, Andreas Fenner, Gregor Henrich, Johannes Kögel, Gabriele Schüler, Irina Greb, Tanja Meth, Burkhard Werner, Joachim Mohaupt, Volker Krau, Matthias Schneider, ACO (Fabrice Bourrigaud, Damien Pourriau), Ferrari S.p.A., Prof. Wolfgang Henseler, Julius Kruta, Mario Linke, Eckhard Schimpf, Frank Stern

1st edition 2024

ISBN 978-3-613-04670-2

DEMAG
CAPITOL
SCHAEFFLER
MICHELIN
POTENZA
BRIDGESTONE
TIM
Shell
BITTE NICHT BETRETEN
DO NOT CROSS